全球低耗材精品住宅设计

田砚杰 主编　　凤凰空间 译

江苏凤凰科学技术出版社

图书在版编目（ＣＩＰ）数据

全球低耗材精品住宅设计 / 田砚杰主编. -- 南京 ：
江苏凤凰科学技术出版社，2014.9
ISBN 978-7-5537-3303-6

Ⅰ．①全… Ⅱ．①田… Ⅲ．①住宅－建筑设计－世界
－图集 Ⅳ．①TU241-64

中国版本图书馆CIP数据核字(2014)第117611号

全球低耗材精品住宅设计

主 编	田砚杰	
译 者	凤凰空间	
项 目 策 划	凤凰空间	
责 任 编 辑	刘屹立	

出 版 发 行	凤凰出版传媒股份有限公司
	江苏凤凰科学技术出版社
出 版 社 地 址	南京市湖南路1号A楼，邮编：210009
出 版 社 网 址	http://www.pspress.cn
总 经 销	天津凤凰空间文化传媒有限公司
总 经 销 网 址	http://www.ifengspace.cn
经 销	全国新华书店
印 刷	北京建宏印刷有限公司

开 本	965 mm×1 270 mm 1 / 16
印 张	21
字 数	168 000
版 次	2014年9月第1版
印 次	2014年9月第1次印刷

标 准 书 号	ISBN 978-7-5537-3303-6
定 价	318.00元

图书如有印装质量问题，可随时向销售部调换（电话：022-87893668）。

PREFACE

序言

人类在地球上已存在 200 万年，但从未像近 100 年来对地球伤害如此巨大，以至于人类在困惑还能存在几个 100 年？人类的体质越来越虚弱，欲望却越来越贪婪，不断地在地球蔓延，向自然索取，成为地球之"癌"。

延续地球的生命就是延续人类的生命。生态建筑是缓解"药方"之一，其从本质上说就是节能型建筑，或称高效益的建筑。用美国建筑师富勒的话说就是"少费多用"；德国建筑师英恩霍文说得更具体明确："用较少的投入取得较大的成果，用较少的资源消耗，获得较大的使用价值。

本书收录国外四十余个公寓和住宅设计实例，从材料、立面、颜色、节能四个角度全方位论述了低耗能居住建筑的设计理念，选取的案例造型现代，图片丰富，对建筑设计人员的设计实践具有很好的指导和借鉴意义。

田砚杰

CONTENTS
目录

CONTENTS
目录

ENERGY 节能

MATERIAL

材料

62 APARTMENTS, PARIS

法国巴黎 62 公寓

Architect: Hamonic + Masson
Client: Paris Habitat
Location: Paris, France
Built Area: 5,120 m²
Status: Completed
Units: 62
Photography: Sergio Grazia

设计公司：Hamonic + Masson
客户：Paris Habitat
地点：法国巴黎
建筑面积：5 120 m²
状态：已建成
户数：62
摄影：Sergio Grazia

The original construction cost is € 8,300,000, that means: construction cost in China is

¥ 15,911,435

per unit cost is ¥ 256,636

原建造成本为 8 300 000 欧元，国内建造成本约

¥ 15 911 435 元

国内平均每户造价约 256 636 元。

The project embraces new concepts of living together primarily based on generous outdoor spaces, both private (balconies) and communal (floor area), as well as on an extrapolation of the advantages of detached houses, which have now disappeared forever from Paris – having one's own floor space and thus being rooted in the soil.

本案提倡新的生活理念，以宽敞的户外空间为出发点，包括以阳台为主的私人空间和以楼层为主的共享空间，重新探索独立式住宅的优势（独立式住宅已经在巴黎彻底消失），为住户打造出扎根于大地的私人空间。

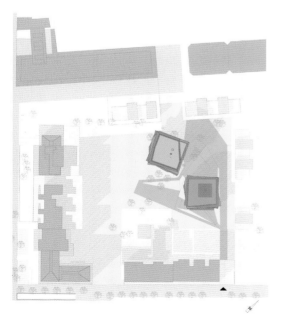

Cost and Creativity 成本与创意

Environmentally friendly materials are extensively used to minimum cost while achieving a maxium sustainable effect.

本案多采用环保型的材料，以最少的造价获得最大的可持续效果。

This in-depth consideration of nature and of the status of the flat's outdoor areas is complemented by a studied choice of materials to clad them, combining different materials and textures and emphasizing color. These arrangements all serve to enrich the façade.

在本案中，设计师还对自然和外部环境作出了更深层次的考虑，通过对材料的研究与整合，结合不同的材质和纹理，强调了颜色的选择，从而极大地丰富了立面的效果。

This project not only offers a living-place but above all a life "on the ground" that lends new meaning to the idea of communal space as an extension of public urban space open to and shared by all. By rooting itself in the city soil, it brings a fresh sense of optimism to a difficult site. The architects hope that this energy will spread from the topmost storeys of the towers to surrounding landmarks such as the Institut Français de la Mode and, of course, the National Library of France, of which our towers form a kind of azimuthal avatar.

本案不仅为居住者提供了生活空间，还提出了"地上生活"的概念，为共享空间的构思增添了新的意义。本案的构思是将共享空间作为城市公共空间的延伸，为大众所共享，并通过与城市之间的紧密关系来弥补场地的不足。设计师希望这个理念可以影响到如巴黎时装学院和法国国家图书馆等周围的地标建筑，引领新的潮流。

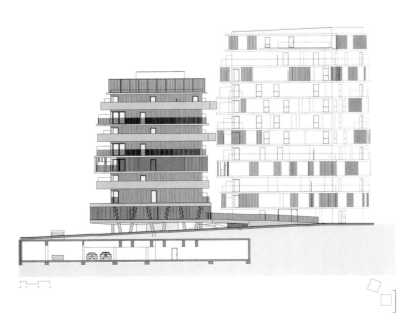

The project involves two blocks of flats, one of 11 storeys above the ground floor and the other of 8 storeys above ground floor. They stand out not only because of their height but also because of their movement, one being a hybrid of the other and their proximity creating the impression of shifting morphology. They are connected by their moving relationship to the ground. A path crosses the block in an arc running from the entrance on rue Villiot to the fire service access on the quai de la Rapée. This walkway, which is covered in a green-colour-red soft material, is bordered by a garden filled with trees and plants.

The project's answer to such a high density of buildings is to be dynamic, go aerial and disrupt the status quo. A new mood is sweeping through landscape and houses, strong enough to lift them off the ground and send them spiraling up into the air. It is this link to the ground that brings coherence to the project. It can be felt on every level and in all aspects of the building, with the green of the balconies (quite thick, like a floating mass) also accentuating this impression that the ground has taken off. This stacking of floor levels defines the architecture of the buildings and the public spaces. The starting point of the project, the ground, accompanies and moulds itself to the natural level, twists and transforms itself, hosts different disciplines, guides and accompanies residents, visitors and passers-by and makes it into a home.

本案共包括两栋公寓楼，分别高11层和8层，不包括地下室。这两栋公寓之所以引人注目不仅因为它们的高度，还因为它们的动感设计。它们彼此之间交相辉映，创造出动感的旋转形态，并以这种旋转的形态在首层相互连通。一条以红、绿色软材作铺装的弧形小路穿过了公寓楼之间，从 Rue Villiot 大道的路口一直通向 quai de la Rapée（巴黎地铁五号线）出口的消防设施处。小路的两侧是种植了茂密草木的花园。

针对高密度的特点，本案采用了"空中旋转"的形式作为应对方案，从而打破了原有的局限。设计师赋予了景观和建筑一种新的风貌，使其看起来仿佛被提升至地表以上，再以旋转的形态缓缓上升，直入云霄，连接了天空与大地，正是这种联系让本案具有了连贯性。从各个角度看，公寓的每层楼都有绿色的阳台，它们看上去像是飘浮在空中的厚重体块，更能凸显出离地而起的感觉。堆叠的楼层界定了建筑和公共空间的范围。本案的起点、地表、景观及其旋转的形态均遵循了不同的原则，为住户、游客和路人呈现出一个温馨的家园。

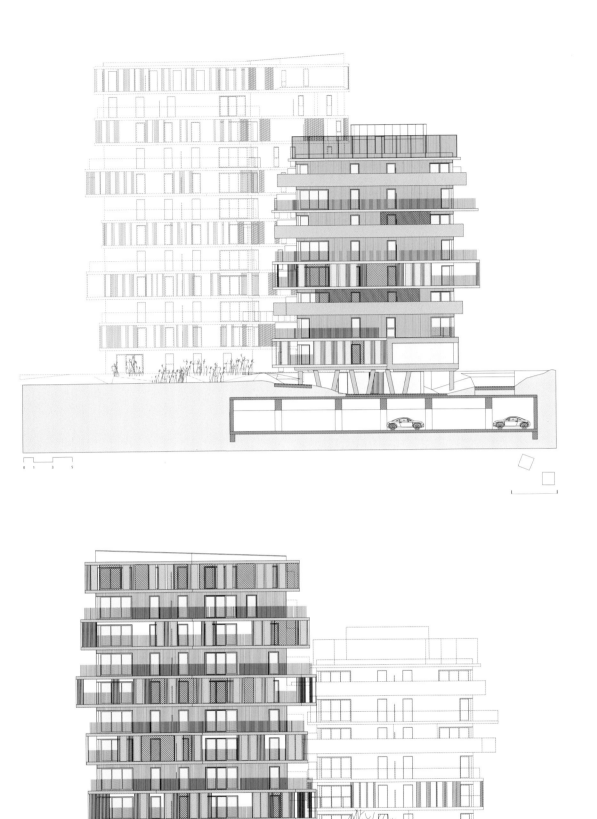

The silver-coloured gangway ceilings underscore the difference between inside and outside. The loggias in the covered ribbons are clad in aluminium and the balcony areas in stainless steel. Then, a system of aluminium screen walls, coloured glass, stainless steel lists and mirror sheets, stacked up on top of each other, storey on storey, contribute to deconstructing the façades and to mix up inside and outside, giving our two towers a Parisian caravanserai look.

过道的银色天花板强调出内部和外部的差异。凉廊以镀铝包裹，而阳台则以不锈钢材料打造而成。铝幕墙、彩色玻璃、不锈钢和镜面一层层地彼此堆叠，打破了立面原有的结构，并将室内、室外空间结合在一起，让两栋公寓楼看起来就像是巴黎旅社一样。

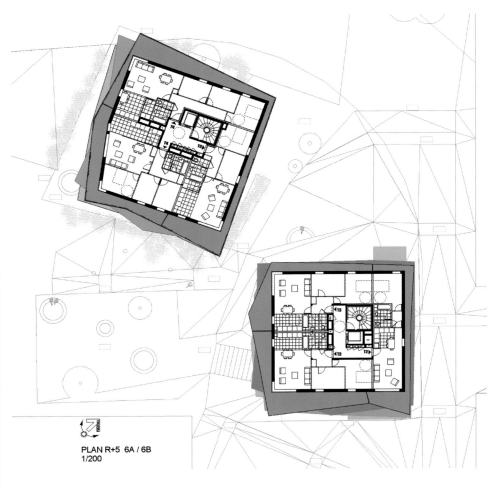

PLAN R+5 6A / 6B
1/200

Each level and each flat has a different floor lending itself to different practices and uses. Rather than being like a balcony, a loggia (or a terrace), which can be seen and used on a daily basis, winds its way around the outside of the flats and gives residents the feeling that they live outdoors. This "poured garden" creates close ties to the building's external environment.

每层楼的每间房都有一片特别的空间，以满足不同的用途和需求。它们既不像阳台，也不像凉廊（或露台）那样常见并只能满足一些日常的用途。这片空间围绕在房间的外部，让住户觉得似乎是在户外活动。这种"户外空间"可以将公寓与其外部环境紧密地联系起来。

MACALLEN BUILDING CONDOMINIUMS

麦卡伦公寓

Architect: Office dA
Client: Pappas Enterprises, Inc.
Location: South Boston, Massachusetts, U.S.A.
Site Area: 32,500 m²
Status: Completed
Units: 140
Photography: John Horner Photography

设计公司 : Office dA
客户 : Pappas Enterprises, Inc.
地点 : 美国马萨诸塞州波士顿南部
占地面积 : 32 500 m²
状态 : 已建成
户数 : 140
摄影 : 约翰·霍纳摄影工作室

The original construction cost is $70,000,000, that means: construction cost in China is

¥ 76,985,449

per unit cost is ¥ 549,896

原建造成本为 70 000 000 美元，
国内建造成本约

¥ 76 985 449 元

国内平均每户造价约 549 896 元。

As a pivotal building in the urban revitalization of South Boston, the Macallen Building's design requires a reassessment of conventional residential typologies to produce an innovative building that works within a developer's financially competitive budget. Occupying a transitional site that mediates between highway off-ramps, an old residential fabric, and an industrial zone, the building negotiates different scales and urban configurations. The design addresses two scales and the different edge conditions of the surrounding context through varied spatial conditions, various ways of reacting to the public sphere, and accompanying material and façade articulations to reinforce the scales of interaction.

作为振兴波士顿南部的关键建筑，麦卡伦公寓的设计要求突破传统公寓的类型，并在开发商的预算内实现创新。公寓位于一片传统型的过渡场地上，周围有高速公路匝道、旧住宅区和工业区，因此须在不同的城市设施中取得协调与平衡。设计师通过多样的空间条件、不同的公共领域处理手法、不同的材料和外墙接合，打造出两个不同的部分来反映城市周边环境，并加强两部分间的互动。

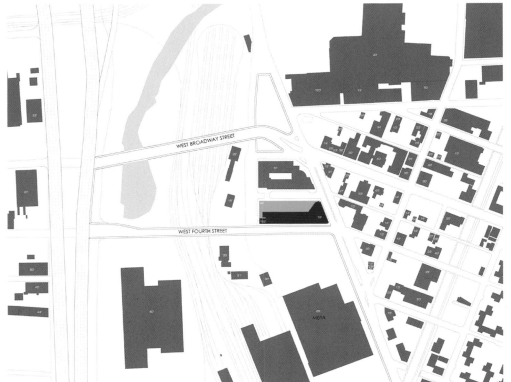

On the western end, the building responds to the highway with a curtain wall yielding panoramic views for the residents inside the building. On the eastern end, brickwork mirrors that of the residential building fabric, extending the logic of the storefront and pedestrian scale elements on that façade. On the north and south façades, bronzed aluminum panels reflect the industrial neighborhood component and express the structural system organization.

公寓的西侧面对高速公路的一面是幕墙，为楼内的居民提供开阔的视野；东侧则是砖砌的立面，将店面和人行道的元素进行扩展；而北侧和南侧的立面则是古铜色的铝板，凸显出毗邻的工业区元素，并表达了建筑系统的组成。

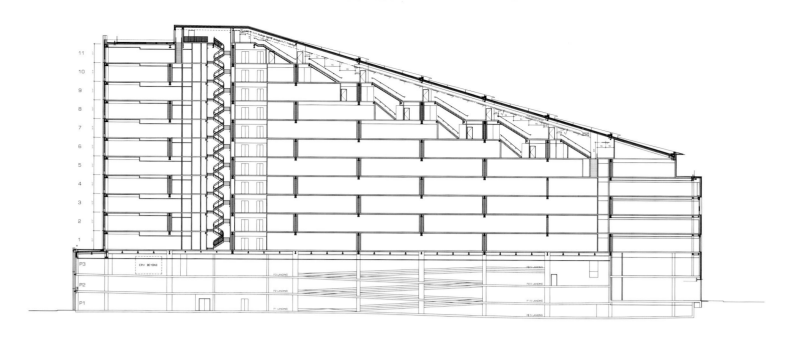

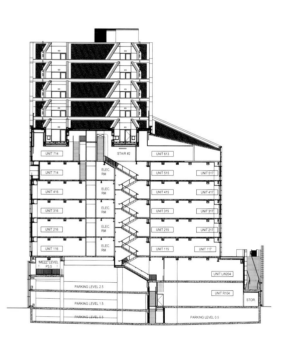

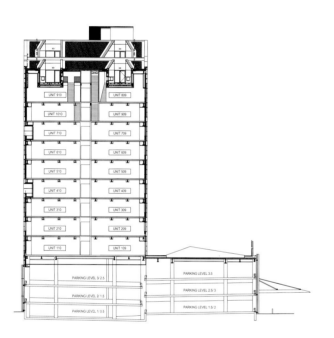

Cost and Creativity 成本与创意

The Macallen Building's design required a reassessment of residential typologies to produce an innovative and sustainable building that worked within a developer's budget. It took advantage of all sustainable building techniques to make it the first LEED Gold certified multi-family housing in Boston.

在本案中，设计师在住宅类型方面做出了新的突破，在开发商的预算范围内建造出创新和可持续发展的建筑，并利用了可持续建筑技术的优势，使其成为波士顿第一个获得 LEED 金牌认证的多户式家庭住宅。

The Macallen building was designed from the ground up to take advantage of "green" building techniques and materials.

从首层开始，本案设计就采用了环保的建筑技术和材料。

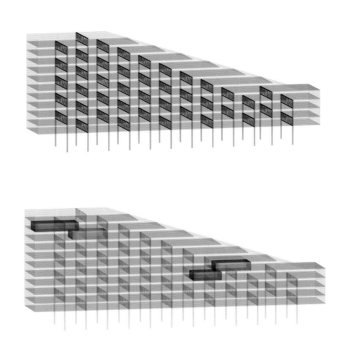

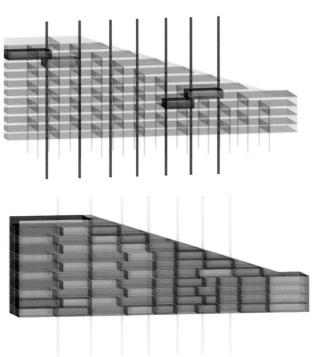

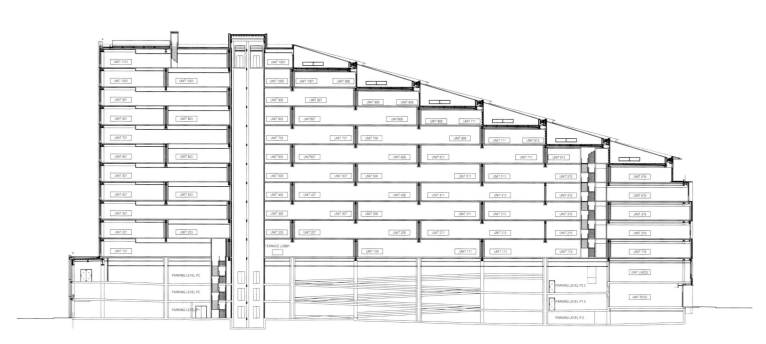

This was a charged and difficult urban site between a historic neighborhood, an industrial wasteland, and an infrastructural interchange. The scale and scope of the project was substantial and demonstrates Office dA's ability to work within the complex process of gaining public approval for a project. This design received LEED Gold certification and is a hallmark of Office dA's experience in advancing and promoting sustainability.

本案所在的场地是一块经过改造的城市用地，其地理条件较为复杂，它处于历史悠久的街区、工业荒地和基建设施的交会处。本案的规模和范围完美地体现出设计公司的能力，并明确无误地告诉人们，该公司在如此复杂的工程中依然游刃有余，得到公众的好评。本案的设计不仅获得了 LEED 金牌认证，同时也是设计公司在推动和促进可持续发展方面逐渐成熟的一大标志。

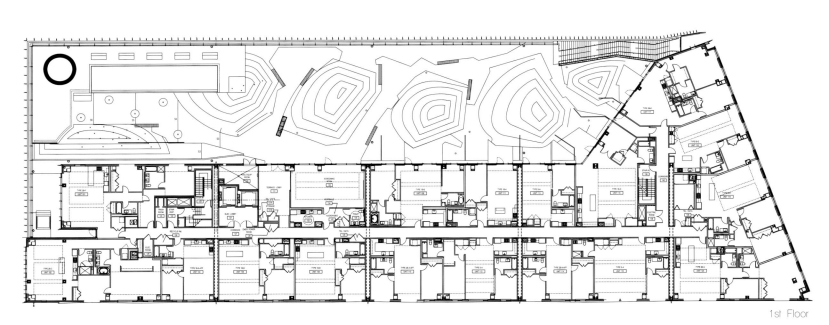

1st Floor

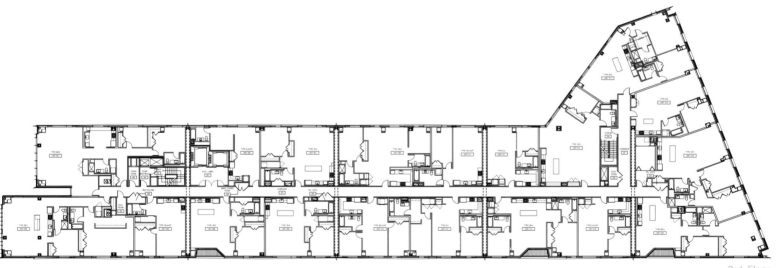

3rd Floor

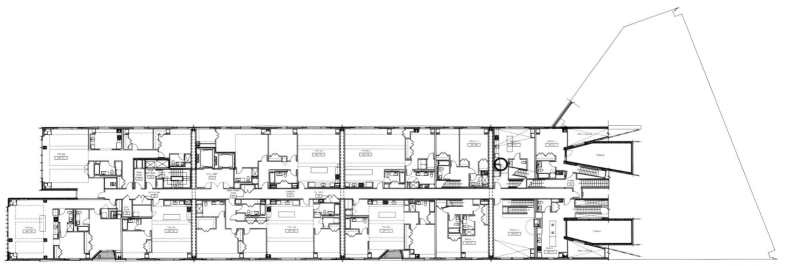

7th Floor

JEWELL

宝石公寓

Architect: Kavellaris Urban Design
Client: Keystone Property Group Pty. Ltd
Location: Brunswick, Victoria, Australia
Site Area: 8,200 m²
Status: Completed
Units: 50
Photography: Peter Bennetts

设计公司：Kavellaris 城市设计事务所
客户：Keystone Property Group Pty. Ltd
地点：澳大利亚维多利亚省布伦瑞克
占地面积：8 200 m²
状态：已建成
户数：50
摄影：彼得·本尼茨

The original construction cost is $12,500,000
that means: construction cost in China is

¥ 16,533,169

per unit cost is ¥ 330,663

原建造成本为 12 500 000 美元，
国内建造成本约

¥ 16 533 169 元

国内平均每户造价约 330 663 元。

The subject site fronts a main street (Union Street) to the south and a private car park that abuts and wraps around the northern and eastern interfaces of the site. The car park service's RMIT University and is wedged between the subject site and a railway corridor. And although the car park is a private space it is utilized as a public thoroughfare from the Jewell train station opposite the University and from Dawson Street to the North. As a result, these site conditions provided for a building that is highly visible without any obstruction.

本案场地的前南方是一条主干道大街（联合大街）和一个私人停车场。停车场毗邻场地东北角，并环绕着这个角落。停车场为墨尔本皇家理工大学所有，处于场地和铁路走道之间。虽然停车场是私人所有，但也被用做公共通道，从宝石车站到对面的大学，从 Dawson 街通向北方。因此，场地周围的环境为公寓提供了开阔的视野。

Cost and Creativity 成本与创意

Simplifying and understanding the construction methodology to execute design.

仔细研究并简化建筑方法，在设计中减少成本。

Three species of timber are used to accent elements of the composition in order to establish an immediate dialogue with the juxtaposed buildings that share the same materiality and also to provide a striking contrast between urban and organic materials. These synergies add to the drama and accentuate the sculptured qualities of the building.

设计师使用三种木材来强调公寓的元素组成，目的是为了与周围并列的建筑建立直接的对话，虽然材料相同，但也能与城市和有机材料之间形成鲜明的对比。这些协同作用增加并突出了建筑的雕塑感。

Legend

1. townhouse unit
2. townhouse backyard
3. communal courtyard
4. carpark
5. public area
6. commercial space

Ground FLoor Plan

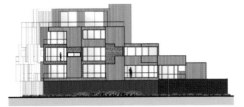

North Elevation

South Elevation

East Elevation

Cross Section

KAVELLARIS architecture interior planning urbandesign consultants
URBAN 53 VICTORIA PARADE, COLLINGWOOD VIC AU 3066
DESIGN T. +61 3 9417 1116 F. +61 3 9417 1119 www.kud.com.au

Our scheme employs passive and active environmentally sustainable technologies and design strategies that were the foundation to the planning of the building. We incorporated landscaped streets and laneways within the subject site to provide spatial separation between dwellings an enable every habitable space to receive natural light and ventilation. In addition, these separations allow the spaces to be naturally cross ventilated thereby minimizing the need for airconditioning. Low emission lighting complimented with solar panels further reduced the carbon footprint of the complex. We also incorporated a 40,000 litre underground storm water collection tank that is used for toilet flushing.

设计师的计划是采用被动性和主动性相结合的环境可持续发展技术和设计战略，这是建筑规划的基础。场地内的街道景观和小巷也被考虑到其中，以划分出每栋建筑的空间，为每户居民提供充足的自然采光和通风。此外，这些分离空间还有助于自然交叉通风，从而最大限度地减少对空调的需求。太阳能电池板的低排放量照明进一步减少了碳的排放。在地下还有一个 40 000 L 的雨水收集罐，收集的雨水可作盥洗用水。

The rhythmic boxed formed geometries are referential variations of the same architectural language resulting in a sculptured building that evokes a sense of movement. As the building is experienced from the various vantage points, the building form takes on different readings to people in trains, cars, on bicycles or walking pedestrians. This movement is reinforced by the reflective nature of the facades metallic cladding that changes the textural qualities and colour of the building with the varying light and shadows of the day.

The asymmetrical composition of slipping corners that wrap around the northern and southern edges of the building blurs the boundaries from the converging facades into a continuous architectural expression. We studied the column motifs of Robert Venturi's design for the Sainsbury Wing at the Nation Gallery in London and incorporated the manner in which the columns negotiated the corner intersections into our design. The southern elevation has been specifically design with a constructed view of the building typical of the classical order when viewed from Watson Street. Vertical strip windows capture the opposing perspective providing a visual connection and a conclusion to the journey for occupants that have entered the building from Watson Street. The accented double height entry void that interrupts the rhythmic articulation of the composition emphasizes and clearly announces the entry into the building as seen from Watson Street.

具有韵律的堆叠体块组成了整座建筑物，这种雕塑化的风格赋予建筑强烈的动感。因为建筑外形多变，从火车、汽车、自行车和路人的角度看都会出现不一样的视觉效果。反光金属表皮增强了建筑的律动感，光线和阴影的变化不断丰富着表皮的质感。

不对称的转角包裹了建筑南北两边，模糊了住宅与立面之间的界限。设计师研究了伦敦国家美术馆里罗伯特·文丘里设计的"塞恩斯伯里之翼"的圆柱样式，并将之吸收作为公寓的转角交叉点的设计。从 Watson 街上看，公寓南立面的设计较为特别，以典型的古典样式作为建筑构造。垂直的长窗能吸引视线，当住户从 Watson 街走进公寓回家时，能提供视觉上的联系，并宣告旅途的结束。双重高度的入口产生的空间打断了有节奏感的建筑连接处，明确地强调出公寓在 Watson 街的入口处。

Legend

1. townhouse unit
2. townhouse backyard
3. communal courtyard
4. carpark
5. public area
6. commercial space
7. internalised street corridor
8. apartment unit

First Floor Plan

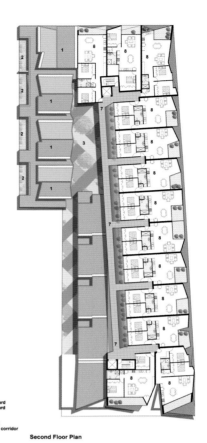

Legend

1. townhouse unit
2. townhouse backyard
3. communal courtyard
4. carpark
5. public area
6. commercial space
7. internalised street corridor
8. apartment unit

Second Floor Plan

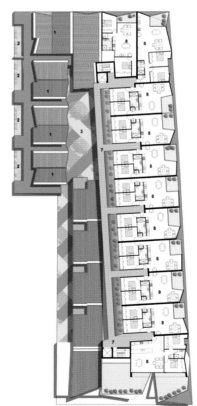

Legend

1. townhouse unit
2. townhouse backyard
3. communal courtyard
4. carpark
5. public area
6. commercial space
7. internalised street corridor
8. apartment unit

Third Floor Plan

Legend

1. townhouse unit
2. townhouse backyard
3. communal courtyard
4. carpark
5. public area
6. commercial space
7. internalised street corridor
8. apartment unit

Fourth Floor Plan

SOCIAL HOUSING TOWER OF 75 UNITS IN EUROPA SQUARE

EUROPA 广场上的 75 房经济公寓

Architect: Roldán + Berengué, arqts.
Location: Barcelona, Spain
Built Area: 10,312.98 m² (Apartment Units: 7,660 m²)
Status: Completed
Units: 75
Photography: Jordi Surroca

设计公司：Roldán + Berengué, arqts.
地点：西班牙巴塞罗那
建筑面积：10 312.98 m²（公寓部分 7 660 m²）
状态：已建成
户数：75
摄影：Jordi Surroca

The original construction cost is €10,083,065 that means: construction cost in China is

¥ 22,188,758

per unit cost is ¥ 295,850

原建造成本为 10 083 065 欧元，
国内建造成本约

¥22 188 758 元

国内平均每户造价约 295 850 元。

The tower E.I.O.5 is a project of social housing promoted by INCASOL and it is located in a new central zone known as Europa Square.

The piece we had built was an object of public client's competition. In our proposal, rescaling the tower according to its position as a piece in the limit with the consolidate fabric of Hospitalet trying to visualize with the building a movement between EUROPA Square and the blocks of 5 floors that form the surrounding.

E.I.O.5 公寓是由 INCASOL 公司进行开发的社会经济住房，坐落在城市的新中心区——EUROPA 广场。

设计公司在公共竞赛中取得优胜，从而承接了本案的设计。在提案中，设计师根据公寓的位置，重新调整公寓的尺寸，并希望在有限的空间里呼应 Hospitalet 市统一的城市环境，并以动态的外表在 EUROPA 广场和周围 5 层高的大楼之间形成别样的视觉效果。

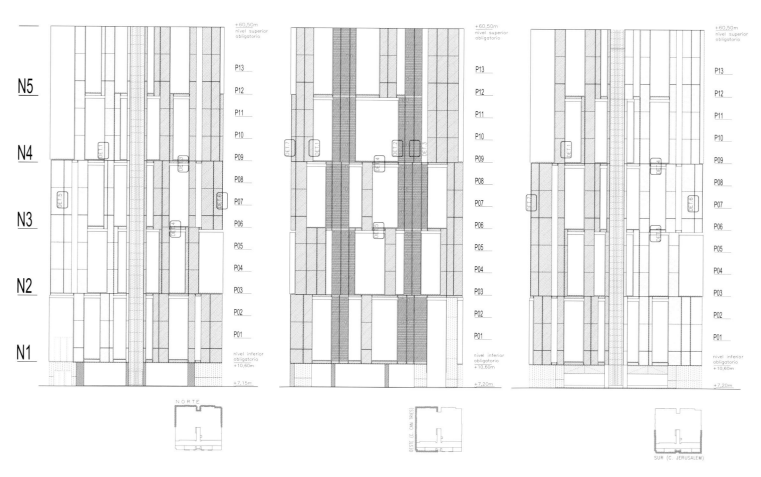

N5
N4
N3
N2
N1

+60,50m
nivel superior
obligatorio

P13
P12
P11
P10
P09
P08
P07
P06
P05
P04
P03
P02
P01

nivel inferior
obligatorio
+10,60m

+7,15m

NORTE

+60,50m
nivel superior
obligatorio

P13
P12
P11
P10
P09
P08
P07
P06
P05
P04
P03
P02
P01

nivel inferior
obligatorio
+10,60m

+7,20m

OEST (C. CAN TRES)

+60,50m
nivel superior
obligatorio

P13
P12
P11
P10
P09
P08
P07
P06
P05
P04
P03
P02
P01

nivel inferior
obligatorio
+10,60m

+7,20m

SUR (C. JERUSALEM)

Elevations

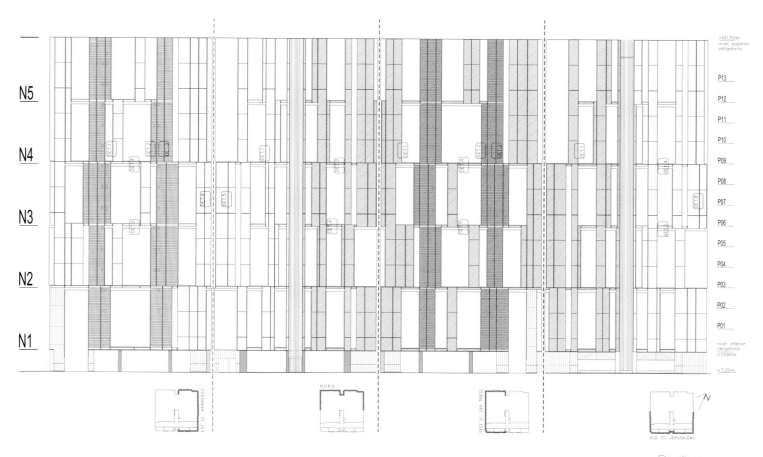

N5
N4
N3
N2
N1

+60,50m
nivel superior
obligatorio

P13
P12
P11
P10
P09
P08
P07
P06
P05
P04
P03
P02
P01

nivel inferior
obligatorio
+10,60m

+7,20m

EST (C. HERRERO) NORD OEST (C. CAN TRES) SUD (C. JERUSALEM)

Elevations

049

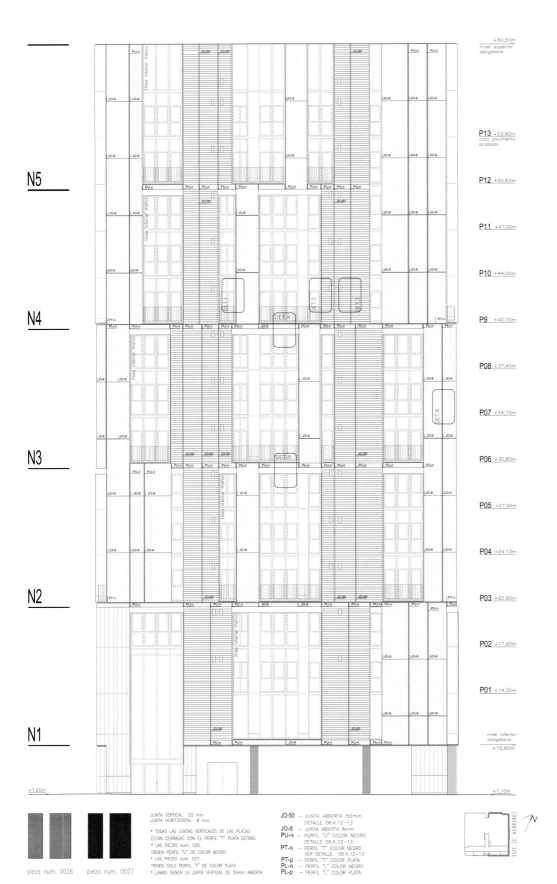

N5

N4

N3

N2

N1

+7,20m

+60,50m
nivel superior
obligatorio

P13 +53,90m
cota pavimento
acabado

P12 +50,60m

P11 +47,30m

P10 +44,00m

P9 +40,70m

P08 +37,40m

P07 +34,10m

P06 +30,80m

P05 +27,50m

P04 +24,13m

P03 +20,90m

P02 +17,60m

P01 +14,30m

nivel inferior
obligatorio
+10,60m

+7,10m

JUNTA VERTICAL 50 mm
JUNTA HORITZONTAL 8 mm

* TODAS LAS JUNTAS VERTICALES DE LAS PLACAS
ESTÁN CERRADAS CON EL PERFIL "T" PLATA DETRÁS
* LAS PIEZAS num. 026.
TIENEN PERFIL "U" DE COLOR NEGRO
* LAS PIEZAS num. 027.
TIENEN SOLO PERFIL "T" DE COLOR PLATA
* LAMAS TIENEN LA JUNTA VERTICAL DE 50mm ABIERTA

pieza num. 0026   pieza num. 0027

JO-50 – JUNTA ABIERTA 50mm
DETALLE 06.K.12-13
JO-8 – JUNTA ABIERTA 8mm
PU-n – PERFIL "U" COLOR NEGRO
DETALLE 06.K.12-13
PT-n – PERFIL "T" COLOR NEGRO
VER DETALLE 06.K.12-13
PT-p – PERFIL "T" COLOR PLATA
PL-n – PERFIL "L" COLOR NEGRO
PL-p – PERFIL "L" COLOR PLATA

East Elevation

Cost and Creativity 成本与创意

The election of the elements for the façade's assembly has been done on evaluating the material price and loss rate, its natural origin and the capacity for being recycled at the end of its useful life.

项目的立面多选用天然和可回收材料，可降低造价以及施工过程中的损耗率。

As some of the images in a "short plane" we have done for the competition show us, these frames avoid vertigo impression because between the interior and the outside of the apartment there is always an intermediate element: balconies, jambs, or the lintels of this big holes.

公寓以短块材料拼接的形式呈现，内部和外部穿插着各种过渡元素，如阳台、侧板和大过梁等，以免人们注视过久后出现眩晕的感觉。

Materially, the façade is constructed with an 8 mm thick HPL pannels hanging of hidden structure of recycled aluminium perfiles. Black frames are made of 4 mm thick aluminium composite panels which brings equal resistance with a lower weight per m² to any other material with the same features.

Synthetically, all materials used in the tower's construction are 100% recyclable and specifically the ones used in the façade come from 65% and 100% already recycled materials.

从材料上说，立面是用 8 mm 厚的 HPL 板构成。黑色的框架是 4 mm 厚的铝制复合板，这种材料与其他具有同等功能的材料相比更为轻薄，并且承重量相当。

公寓使用的所有材料都是百分百可回收的，特别是立面，由 65% ~ 100% 的回收材料打造而成。

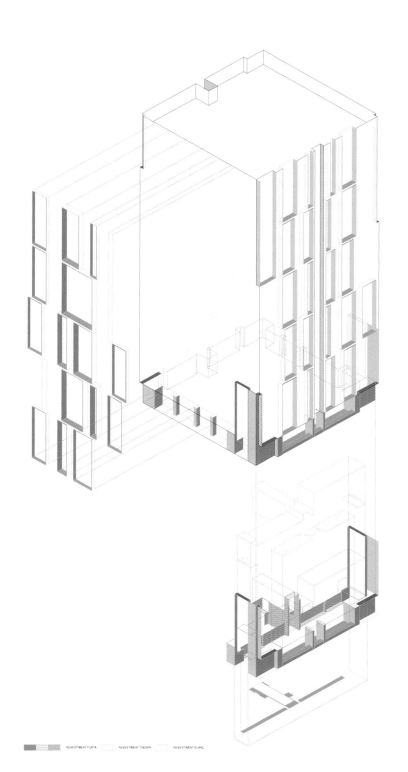

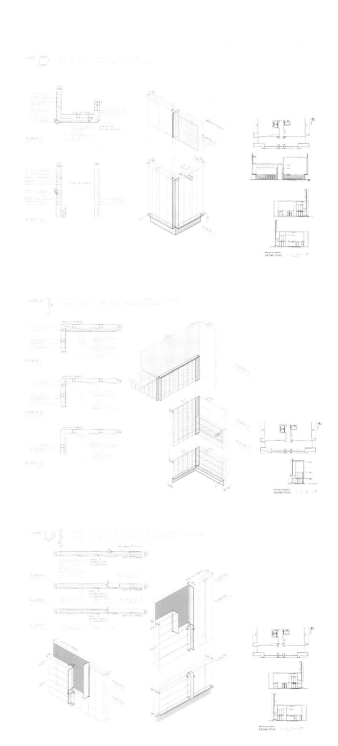

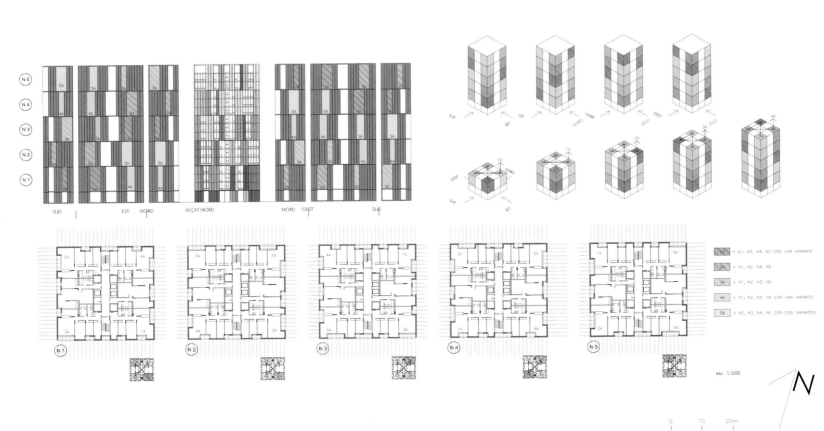

SUD   EST NORD   ALÇAT NORD   NORD OEST   SUD

N 5   N 4   N 3   N 2   N 1

N 1   N 2   N 3   N 4   N 5

1A > N1, N3, N4, N5 CON UNA VARIANTE
2A > N1, N2, N4, N5
3A > N1, N2, N3, N5
4A > N1, N2, N3, N4 CON UNA VARIANTE
5A > N2, N3, N4, N5 CON DOS VARIANTES

esc. 1/500

0   10   20m

N

The tower is set in floor plan as two towers with T form circulation corridor and two scales in the extremes across the one illuminates the interior in the circulation zone. Every tower has two apartments of 69 m² in the extremes and one of 56 m² in the central position. Totally, the program is of 75 units.

In this Project, the apartment unit's distribution begins in the 14th floor and keeps descending so the compacted residential volume from the superior floor releases a space of 3 levels height with a T form in the access. This piece, generous in volume and "sober" in measures and finishing materials, in its longest dimension it ends being a street with accesses in the extremes and has unusual dimensions for the building's entrance hall of usual public housing buildings. We think that big frames of the facades as well as the space of the entrance hall, work like intermediate spaces of relation, shaping the community scale, between the individual and private scale of the houses and the public city scale.

公寓由两栋塔楼构成，以"T"形的走廊连接，中间为社交区。端头户型为 69 m²，中部户型为 56 m²。整栋楼一共 75 个单位。

公寓房间位于 14 层以上，并开始不断下降，一直到贴紧公寓的上层，形成一个 3 层高的"T"形空间作为入口。 公寓通过庞大的体量、精准的尺度和饰面材料，在纵向连接着街道和入口，让公寓的门厅和其他公共建筑相比有着不一样的规模。设计师认为，立面的大型框架和入口门厅其实是不同空间交接的中介，既塑造出社区空间的气质，又能在私人和公共中保持平衡。

TETRIS APARTMENTS

俄罗斯方块公寓

Architect: OFIS Arhitekti
Client: Gradis G Group
Location: Poljane 2, Ljubljana, Slovenia
Site Area: 5,500 m²
Status: Completed
Units: 56
Photography: Tomaz Gregoric

设计公司：OFIS Arhitekti
客户：Gradis G Group
地点：斯洛文尼亚卢布尔雅那市 Poljane 2
占地面积：5 500 m²
状态：已建成
户数：56
摄影：Tomaz Gregoric

The original construction cost is€3,280,200, that means: construction cost in China is

¥ 7,718,407

per unit cost is ¥ 137,829

原建造成本为 3 280 200 欧元，国内建造成本约

¥7 718 407元

国内平均每户造价约 137 829 元。

The location of the site is on the edge of the planned 650 apartments which were finished a year earlier. However, these apartments differ from the previous as they were planned as social housing and were sold to the Slovenian Housing Fund.

公寓坐落于去年完工的 650 号公寓旁，和先前建成的公寓楼不同，此地的公寓楼都将作为社会经济住房，归斯洛文尼亚基金会所有。

Cost and Creativity 成本与创意

The building has a very good ratio between apartment and common spaces, surface used for staircases and other communication. Materials used are affordable but well made, the buildings' structure is rational and organized so that floor plans are flexible, as the structural walls in the building are used to create only the shell of each apartment while interior walls are non-structural.

公寓、公共空间楼梯和其他交流区等之间的分布十分合理。选用的材料物美价廉，公寓的整体结构合理，布局灵活。结构墙只是用来分隔每间公寓，内墙则可以灵活多变。

Tetris Apartment is made of economic but quality materials.

俄罗斯方块公寓的选材兼具了经济性和高品质的特点。

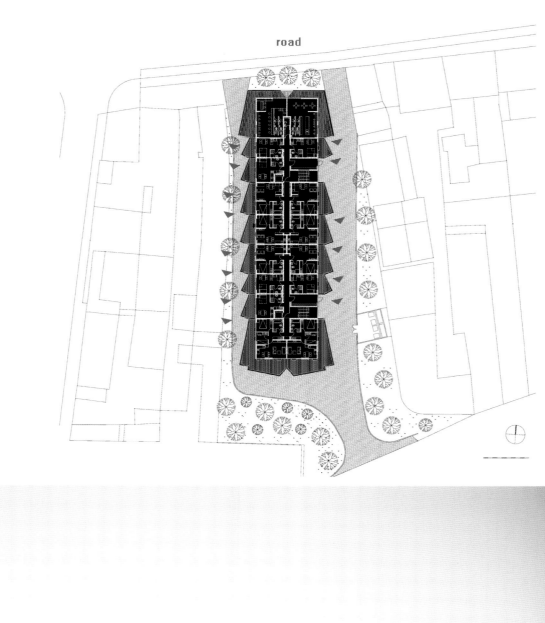

road

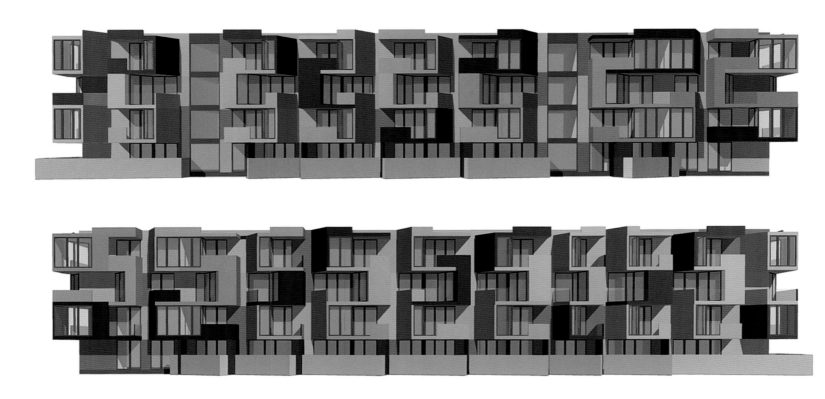

The given urban plot of the building was 4 floors high, 58 meters long and 15 in width. Since the orientation of the building is towards the busy highway the apartment openings, together with balconies were shifted to become 30 degrees window-wings towards the quieter and southern orientated side. In the future 2 additional blocks are planned on both longitudinal sides; therefore there are no direct windows towards east and west. Each apartment has a view towards its own balcony, sometimes there is also a glazed loggia; it creates a feeling of intimacy. There will also be no direct views from ones apartment directly to the others in the opposite block. The apartments are of different sizes – from 30 m^2 studio flats up to 3 room apartment of 70 m^2. Bigger apartments are developed on the front facades and have nicer views and a corner orientation.

公寓的规划空间范围是高 4 层，长 58 m，宽 15 m。由于公寓的规划朝向繁忙的高速公路，因此公寓的开口处和阳台都设计成 30°的倾斜玻璃墙，以更多地朝向安静的东南方向。因为不久后公寓附近还会再建设两栋建筑，所以没有设置朝南和朝北的窗口。公寓里的每间房都有视野良好的阳台，部分房间还有玻璃凉廊，为用户带来私密感。公寓的设计也别出心裁，避免了对面楼宇的视线直视。公寓内的房间有不同的尺寸，有小到 30 m^2 的工作室或大到 70 m^2 的 3 室公寓，还有面积更大的套房设置在公寓楼的正面，具有更好的视野和方位优势。

They are made of economic but quality materials such as wooden oak floors, granite tiled bathrooms and have large windows with external metal blinds. The concept of the structure is made in such a way, that floor plans are flexible, since the only structural walls in the building are used to create the shell of each apartment. All other inner walls are non-structural. Long after the elevations were planned many people associated them to the game Tetris and so the building got its name. The façade was developed simply – just by tracing the floor plan organization. The inner – structural façade wall has plaster; the external wall that embraces the loggia is glazed or wrapped into precast panel. These panels are wooden of three colors that are tracing the vertical zigzag pattern. Balcony fences are either a perforated precast panel or transparent metal fence.

虽然是经济型住房，但设计师选择的都是优质材料，如橡木地板、花岗岩地砖浴室和外部的金属百叶窗等。公寓的结构设计也很巧妙，因为公寓的固定结构的外墙能起到遮风挡雨的作用，所以公寓楼层的布局都各不相同，由此决定了其他内置墙也无固定结构。建成后公寓的立面让人们联想到了俄罗斯方块游戏，公寓也由此得名。公寓的立面规划其实很简单——仅仅是跟着公寓楼层的形状做相应的调整。立面的内墙用石灰涂面，外墙上安置了玻璃或预制板材的凉廊。木制的板材有三种颜色，并刻上了垂直的锯齿状图样。阳台的护栏用的是穿孔预制板或透明的金属护栏。

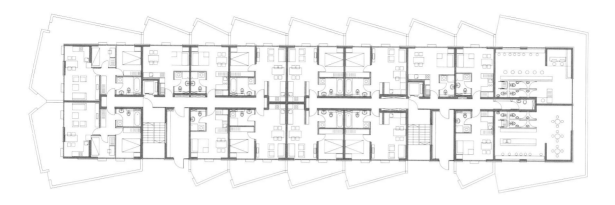

Ground Floor

3rd Floor

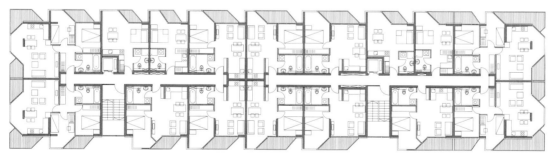

2nd Floor

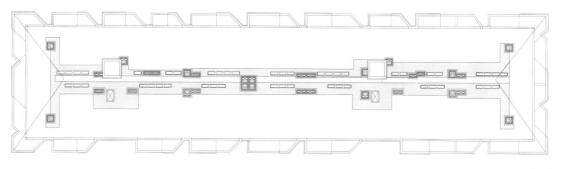

Roof

EL NODO HOUSING

EL NODO 住宅楼

Architect: Exit Architects
Client: SOGEPSA. PRINCIPADO DE ASTURIAS
Location: Avilés, Spain
Built Area: 2,737 m²
Status: Completed
Units: 200
Photography: Miguel De Guzman, Exit Architects

设计公司：Exit 建筑事务所
客户：SOGEPSA. PRINCIPADO DE ASTURIAS
地点：西班牙 Avilés 市
建筑面积：2 737 m²
状态：已建成
户数：200
摄影：Miguel De Guzman, Exit 建筑事务所

The original construction cost is €2,233,087 that means: construction cost in China is

¥ 4,914,124

per unit cost is ¥ 24,571

原建造成本为 2 233 087 欧元，国内建造成本约

¥4 914 124 元

国内平均每户造价约 24 571 元。

From the beginning of the project we considered the quality of the surroundings and wanted to treat the buildings as if they were almost "alive" and could express the sensations that the place provoked in ourselves.

设计师在项目设计之初就考虑周围环境的品质，希望将这些建筑物当做有生命的事物来打造，能够表现这个地方带给我们的震撼。

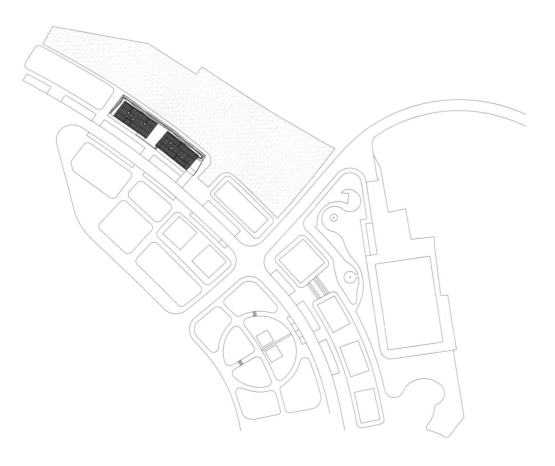

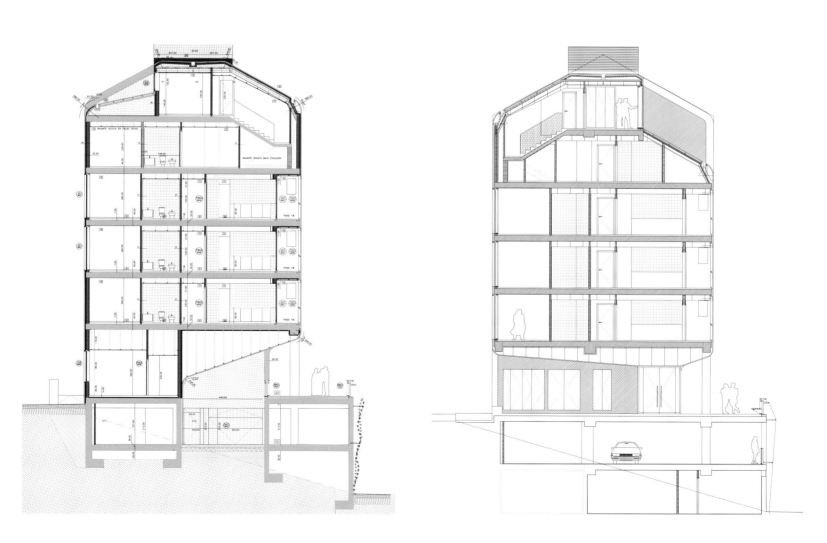

Cost and Creativity 成本与创意

About the control of budget, a very complete implementation project, where all possible contingencies have been taken into account. There may be changes, but they are always acceptable without deviation of budget.

在成本控制方面，在项目完全竣工前，设计师就预先考虑了所有可能发生的意外。即使发生了意外情况，也是在成本的接受范围内。

About the materials, it is all about also simplicity and efficient.

建筑用料力求简单、高效。

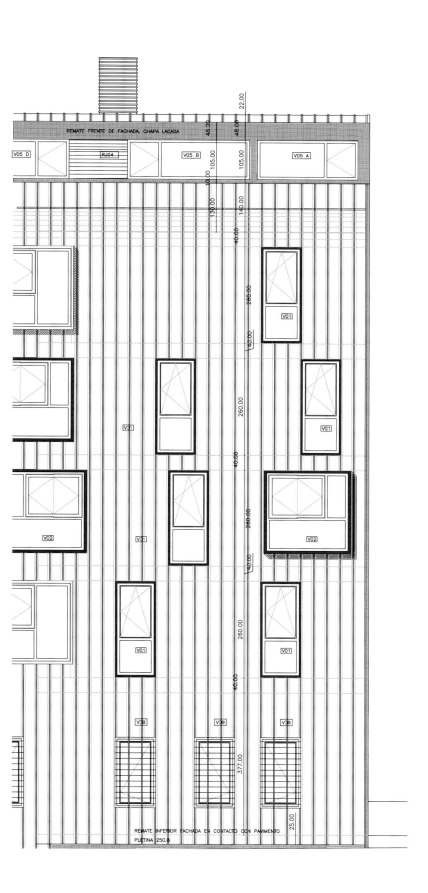

In a privileged place, at the end of a green meadow on the top of a hill and with the Avilés estuary on the horizon, two small metallic objects stand still upon a concrete basement and stare at the spectacle before their eyes. Dressed in a tailored suit which covers their head and torso, they protect themselves from the unpleasant rain and see the time going by.

在山顶上绿地的尽头、地平线上的 Avilés 河口，两栋小金属建筑矗立在水泥基座上，仿佛瞪眼看着眼前精彩的一切。金属建筑体的顶部和主体覆以量身定制的表皮，以免受雨水侵蚀，任时光悠悠流过。

The project idea is all about the covering, the skin, which folds to configure the building personality. In a city with a strong industrial character, we found steel very appropriate, as a tight sheet that gets torn to create windows and balconies, to open itself at the top to the surroundings through big windows that look, as eyes, at the meadows and estuary.

项目的全部理念在于表皮，它们互相交叠，形成了建筑物的个性。在一个重工业城市，钢铁制品非常适合用来包覆窗户和阳台。站在能环顾四周的高处，通过像眼睛一样的窗户就可以遥望远处的牧场与河流。

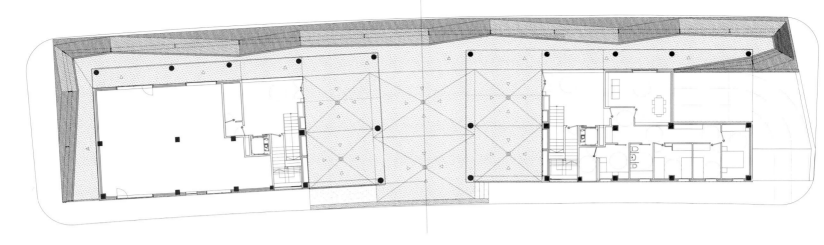

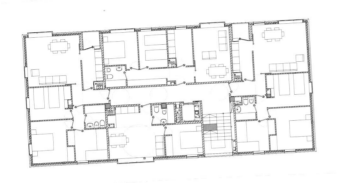

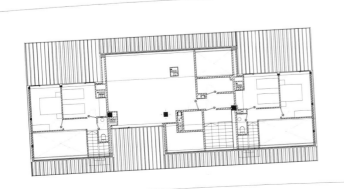
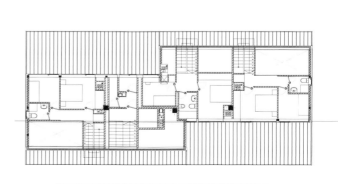

We designed an only and common access to both blocks just in the space between them, with the tension that provoke two things that are very close each other, and as a meeting place for the neighbours. The access level is slightly higher than the street and conceived as a semipublic square where people can stare at the landscape or talk sitting in the benches protected from the rain.

Each block has a staircase whose appearance from the exterior is minimized by using the same steel sheet, but perforated. During the day it looks as the rest of the façade but, at night, light comes out through the steel and its presence becomes more important.

The buildings get anchored in the ground by a concrete basement which contains the parking and lumber rooms and forms the access platform. Upon it settles itself a light metallic skeleton that allows the façade flexibility. The perimeter pillars get reduced to fix into the outer wall and allow the free position of the inner partitions.

Steel sheet with a special sea-climate-proof coat, and outdoor Aquapanel for the side facades. The outside pavement is made of washed concrete while the perimeter fence is solved with a double simple-torsion galvanized net.

设计师在两座大楼之间空地上设计了唯一的公共入口，使它们更加靠近，入口处更像是居民们开会的地方，它比街道略高少许，也像是半公共化的场所，人们可以在这里凝视风景，闲时坐着聊聊天或避雨。

每栋大楼都有一座楼梯，使用带孔的薄钢板，外形紧凑。白天，它展示出简洁的外观；晚上，灯光从钢板孔中泻出，衬托出其优美的形状。

建筑物牢固地耸立在地面，底下是钢筋混凝土基础，里面是停车场、储藏室，形成了通道平台。地上部分是轻钢骨架，使正立面显得更柔和，支柱缩入外墙使内部分区空间更大。

外立面使用海洋气候专用的防潮薄钢板，外部的人行道使用水磨混凝土，边界围栏使用了镀锌钢丝网。

STUDENT RESIDENCE, PARIS

巴黎学生公寓

Architect: LAN Architecture
Location: Paris, France
Site Area: 3,950 m²
Status: Completed
Units: 143
Photography: Julien Lanoo

设计公司：LAN 建筑事务所
地点：法国巴黎
占地面积：3 950 m²
状态：已建成
户数：143
摄影：Julien Lanoo

The original construction cost is €8,000,000
that means: construction cost in China is

¥ 15,336,323

per unit cost is ¥ 107,247

原建造成本为 8 000 000 欧元，
国内建造成本约

¥15 336 323 元

国内平均每户造价约 107 247 元。

The project for a student residence was considered in the context of the urban fabric of the La Chapelle district in Paris and its role in its evolution. The plot is on the corner of rue Philippe de Girard and rue Pajol in the 18th arrondissement, close to the ZAC Pajol, an ambitious redevelopment of former railway yards, on which social, cultural and sports amenities are currently being created. The district is a very heterogeneous mixture of Haussmannian residential buildings, factories and workshops, and therefore has a richness and wide diversity of situations unusual within Paris itself.

本案被看做是巴黎拉夏贝尔区城市结构的演变，位于第 18 郡的菲利吉拉德和帕吉尔地块的一角，紧邻帕吉尔，是个旧铁路货场的再发展项目，生活服务、文化和体育设施正在这里兴建。该地区有不同类型的建筑，包括奥斯曼式住宅楼、工厂、车间，体现出于它所在的巴黎鲜有的丰富性和多样性。

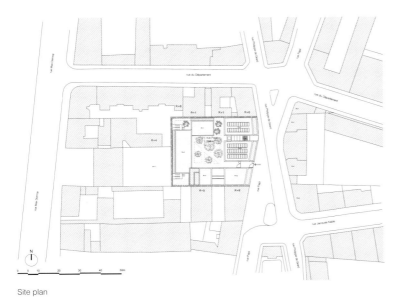

Site plan

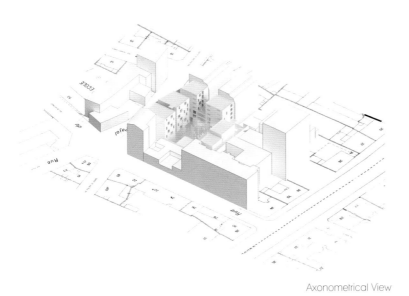

Axonometrical View

Cost and Creativity 成本与创意

Empty spaces in the courtyard and circulations could might be defined as "junkspaces" as they are by-products of the design of the buildings. Our ambition was to restore the value of these spaces in aim to create a genuine "plus" for the residents.

庭院和走道中的每个小空间都被定义为"再生空间"，因为它们是公寓设计中的副产品。设计师希望能充分利用这些空间，为住户创造一个名副其实的"附加空间"。

The choice of materials was dictated by technical and architectural concerns. Our research was guided by a desire for durability and the sober, refined and classical nature of our project.

材料的选择根据建筑和技术需求而定，为实现项目的持久、清洁、典雅、自然而服务。

The driving idea guiding our project stems from the challenge of responding to the necessity for urban integration and creating optimum comfort for the residence's occupants in a convivial and intimate environment.

项目的设计理念来自于城市居住者对轻松私密环境和舒适安逸生活的追求。

0 5 10m

Section AA

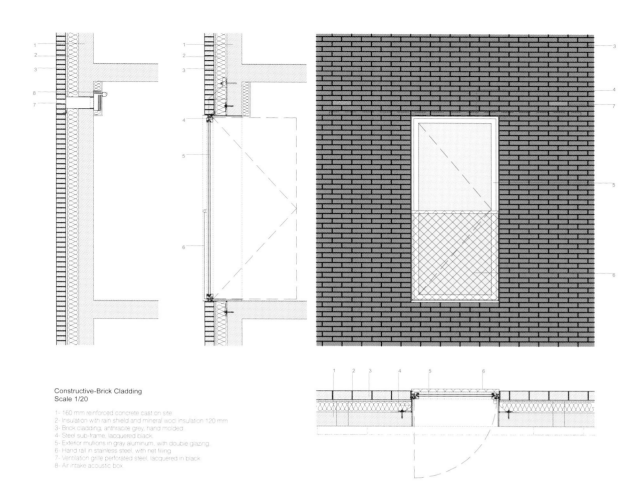

Constructive-Brick Cladding
Scale 1/20

1- 160 mm reinforced concrete cast on site
2- Insulation with rain shield and mineral wool insulation 120 mm
3- Brick cladding, anthracite grey, hand molded
4- Steel sub-frame, lacquered black.
5- Exterior mullions in gray aluminum, with double glazing.
6- Hand rail in stainless steel, with net filing
7- Ventilation grille perforated steel, lacquered in black
8- Air intake acoustic box

The strategy of creating a duality between street and courtyard was pursued in the choice of materials. The facades, instead of imposing a single image on the project, participate in creating the varying atmospheres of the spaces they envelop and delimit. The buildings on the street are clad in dark, slate-coloured brick, while the buildings around the courtyard are clad with larch planking with folding louvred shutters in front of the windows and balconies. The facade along the entry passage is also clad with larch and announces the feeling of the space within. All the ground and wall surfaces in the courtyard are clad with the same light-coloured, flexible material, normally used for sports areas and playgrounds.

创建街道和庭院间二元性的关键是材料选择，立面在空间氛围不断变化中由单一到多样，临街的建筑都使用黑色和灰色的砖，临院的建筑窗户、阳台都装有松木百叶窗，沿入口处通道的立面也装有松木百叶窗以保持与空间的感觉一致，院子里所有墙壁和地面都采用相同的浅色柔性材料，这些材料通常用于体育馆和游乐场。

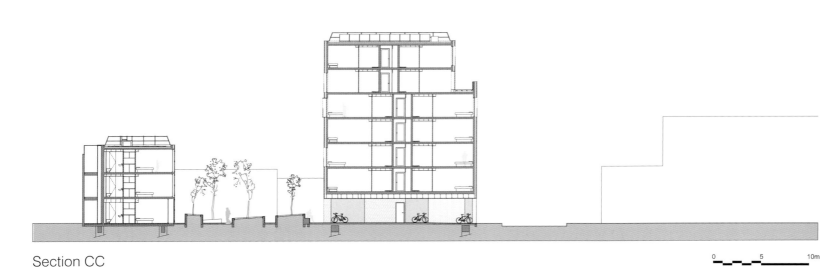

Section CC

0   5   10m

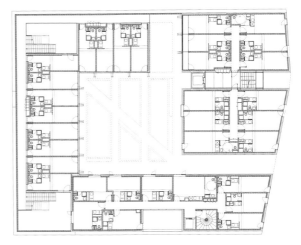

1 Entrance corridor
2 Administration space
3 Garbage storage
4 Hall
5 Laundry
6 Public toilet
7 Informatic room
8 Bicycle parking
9 Guardian house

N

0   5   10m

Ground Floor

First Floor

The project is composed of several buildings, whose volumes and voids depend on the context. On the street, three six-storey volumes are separated by two rifts providing access to the residence and vertical circulation. The heights of the buildings at the back of the plot vary according to neighbouring buildings. In the middle, a spacious courtyard is lit by a rift in the south building, an extension of an existing void.

The courtyard, the heart of the project, provides access to the various buildings and defines their interrelationship. A 15 ×15 metre square, it ensures sunlight for all the rooms and acts as a kind of green lung.

本案由数座建筑组成,其体积取决于环境的大小,三栋六层楼房之间的窄巷,是通向住所的通道和活动场所。地块后部建筑物的高度取决于邻近的建筑,中间是个宽敞的院子,光线从南面建筑物缝隙中照进来,延展了空间。

院子是本案的核心,它提供了联系通道,界定了空间的相互关系,一块 15 m×15 m 的绿地既确保了每间房的光照,也是小区的"绿肺"。

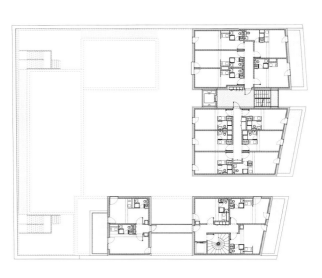

Fifth Floor

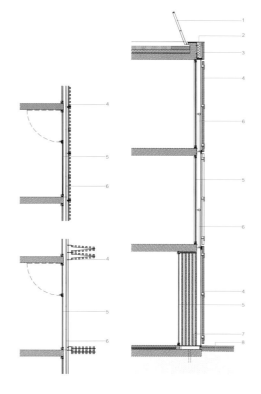

Detail Shutters
Scale 1/50

1- Roof handrail painted black.
2- Aluminum capping sheet painted black.
3- Fixed panel with perforated wood cladding
4- Foldable shutters perforated wood cladded
5- Exterior mullions in gray aluminium, with double glazing
6- Stainless steel rods on stainless steel frame handrail
7- Wood deck
8- Soft floor on recycled tires with rubber granules coating

The project complies with the "Habitat and Environment" label's VHEP specifications. A combination of compactness, treatment of the envelope, and solar heat coupled with high-performance ventilation and heating (urban heating and solar panels) creates pleasant and comfortable accommodation. The concrete structure, insulated on the outside with 12 cm of mineral wool, brick or wood cladding and high-performance double-glazed fittings, provides efficient thermal insulation. In winter the buildings retain their interior heat, and in summer their exterior insulation reduces solar and internal overheating, while inertia enables the capture of daytime heat and its retention during the night.

本案在能源上符合了人居与环境的 VHEP 标准，利用太阳能发电，同时提供高性能的通风和供暖设施，形成令人愉快的住宿环境。混凝土结构外采用 12 cm 厚的矿棉、砖、木包层和双层玻璃表皮作为保温层，隔热性能良好。冬天建筑可以保留其内部的热量，夏季则可以降低紫外线的照射，防止内部过热，而惯性作用则可以将白天得到的热量保留到夜间。

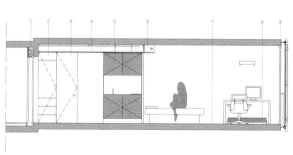

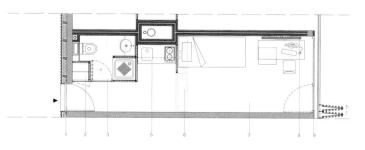

**Constructive Details Plan/Section
Scale 1/50**

1- Metallic entrance door mounted on
sub-frame.
2- Closet with clothes rail and shelving,
painted wooden doors.
3- Bathroom door, opening in painted
wood invisible hinges.
4- Dropped ceiling in polycarbonate,
backlight by neon.
5- Melamine kitchenette in wood white /
black, sink, stove, recessed fridge.
6- Painted wooden shelve, lighting spot.
7- Gray rubber floor.
8- Radiator fins white painted steel.
9- Set of external mullioned, handrail and
shutters.

NURSING HOME IN BAÑOS DE MONTEMAYOR

西班牙 BAÑOS DE MONTEMAYOR 的疗养中心

Architect: GEA Arquitectos
Location: Cáceres, Spain
Built Area: 3,652.60 m²
Status: Completed
Units: 72 Beds (39 bedrooms)
Photography: Ignacio Marqués, Primeros Planos

设计公司：GEA Arquitectos
地点：西班牙卡塞雷斯省
建筑面积：3 652.60 m²
状态：已建成
户数：72 个床位（39 间卧室）
摄影：Ignacio Marqués, Primeros Planos

The original construction cost is €3,000,000 that means: construction cost in China is

¥ 6,601,790

per unit cost is ¥ 169,277

原建造成本为 3 000 000 欧元，
国内建造成本约

¥ 6 601 790 元

国内平均每户造价约 169 277 元。

The project is located in a small town in northern Extremadura called "Baños de Montemayor", surrounded by mountains and forests of chestnut and oak trees. The parcel is in between the urban core, composed of small volumes.

本案位于 Extremadura 区北部的一个名为 "Baños de Montemayor" 的小镇上，周围环绕着群山、栗子林和橡树林。建筑处于城乡间的核心位置，由多栋小型建筑组成。

Cost and Creativity 成本与创意

Both the organization of the work process, and the construction systems used were adapted to the professional abilities of the companies involved in the construction process. We mainly focused on the simplicity of the execution of the works to ensure the participation of local professionals. On the other hand, the systematization of prefabricated wet cells, helped us cut down costs.

施工过程的组织和所使用的建筑系统都邀请了具有相应专业能力的公司参与。设计师较为关注如何简化施工过程，以确保本土的专业人士能参与其中。另一方面，预制的湿电池系统也有助于降低成本。

The materials are the same ones used on the nearby buildings: granite and white plaster. Granite is chosen as material for contacting the ground, looking for its original texture, so that the pure white volumes would emerge cleanly.

建筑采用的材料和附近的建筑相同：花岗岩和白石膏。选择花岗岩作为和地面接触的材料，是因为其原有的质地能与建筑白色的形体形成了鲜明对比。

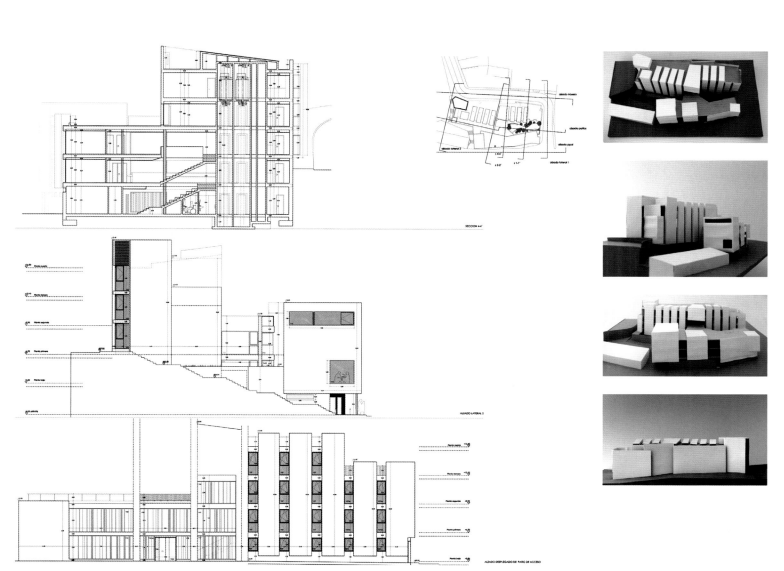

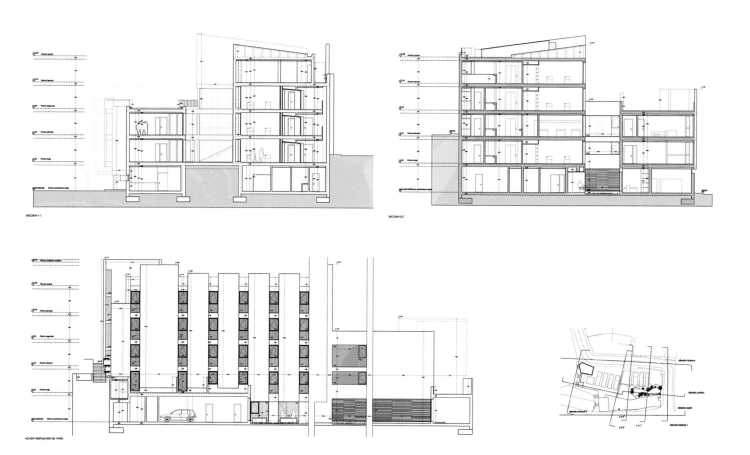

We decided to work on a very fragmented structure and volume looking for a smooth transition. From the top of the mountain, the building offers a clear dialogue between the two scales.

As a constructive approach, each bedroom´s toilet is entirely prefabricated and laid on construction site once the floors are raised, saving costs and time. These concrete cells emerge into the corridors qualifying the access to each bedroom, their color differ on every floor in order to avoid confusion for residents.

设计师决定用一个非常分散的结构和体量模式来寻求平稳的过渡。从山顶上看，建筑与周围环境和谐一致。

建筑采用了预制的方法，每间卧室里的洗手间都是预先建造的，放置在施工现场，等地板铺设完成后就立即搬入，以节约成本和时间。这些独立的单元分布在走廊的每间卧室旁，每层的颜色各不相同，以避免混乱。

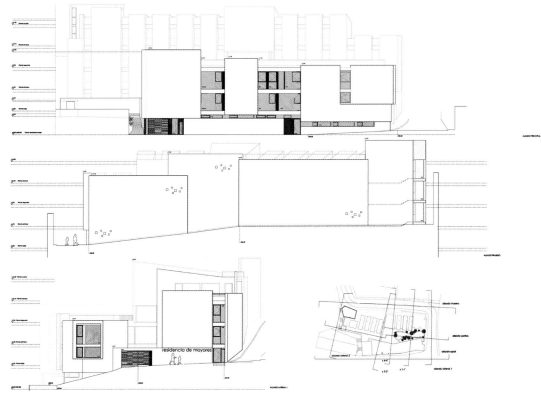

residencia de mayores

The open public spaces (sometimes half opened), separate day and night zones of the building. The sun penetrates over the built volume that consciously is cut where necessary. Everyday life is made there, in those areas stressed that ultimately make up a core of the lot. The surrounding volumes contain some necessary uses: on one side dining, fitness, nursing, cafeteria ... in the other one sleeps.

The bedrooms are all oriented to the west, where the main views are. The corridors are against the mountain and lit by small openings tangent to the path. On the facade of the room's side, the arrangement of recessed windows allows shelter from the midday sun.

开放或半开放的公共空间将建筑的日常活动区和休息区分开。阳光可以透入室内，如果有需要也可以进行遮挡。各种公共空间组合在一起，形成了日常活动区域。周围的小建筑则包括了一些必要的设施：一侧是用餐区、健身房、护理室、食堂等，另一侧则是休息区。

卧室全部朝向西边，视野开阔。走廊面对着群山，光线通过走廊上的小口射入其中，这些小口与小路相切。紧贴立面一侧的房间都设置了凹进的窗口，可以遮挡正午的阳光。

190 HOUSING UNITS IN SALBURUA, SPAIN

西班牙 SALBURUA 区 190 公寓

Architect: Roberto Ercilla Arquitectura
Client: Larcovi S.A.L.
Location: Salburua, Vitoria, Spain
Site Area: 27,715 m²
Status: Completed
Units: 190

设计公司：Roberto Ercilla Arquitectura
客户：Larcovi S.A.L.
地点：西班牙维多利亚市 Salburua 区
占地面积：27 715 m²
状态：已建成
户数：190

The original construction cost is €13,905,000 that means: construction cost in China is

¥ 30,599,296

per unit cost is ¥ 161,049

原建造成本为 13 905 000 欧元，国内建造成本约

¥ 30 599 296 元

国内平均每户造价约 161 049 元。

The proposed solution deals with the urban character of the closed block with a common open space for the community – the inner patio. To provide an adequate scale to the project, the following is proposed.

本案的基本设计构想是运用社区内部公共开发空间——天台——来应对城市建筑物的封闭特性。为实现项目规模，采用如下方案。

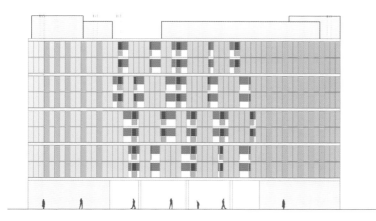

Cost and Creativity 成本与创意

The glass façade contributes not only to low cost, but also to a good daylighting.

由玻璃组成的立面不仅造价低，还能获得充足的自然采光。

Regarding composition, the plans are grouped in twos, giving the façade a greater, less fragmented scale.

整个规划由两大部分构成，尽可能给出一个大外立面，减少零碎感。

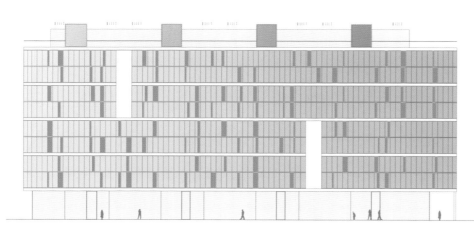

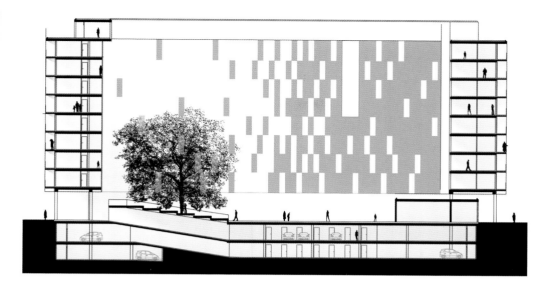

Unitary treatment of the outer façade by means of translucent glass cladding that "dematerializes" the strong presence of the closed block.

The window model, represented in the sliding elements of the exterior façades, is moved to the inner façade. A puzzle of grey-toned pieces is composed based on this model, "pixelizing" the inner façade like a large mural.

The construction system of the cladding, before the glass façade, is made up of continuous insulating cladding that guarantees the absence of thermal bridges.

The incorporation of a large tree in this inner space allows the lighter areas of the puzzle to relate to its presence. Besides the large openings made on façades, the patio opens up on ground floor on the shorter sides, north and south. Stands over the car ramp provide the image of public space.

建筑外墙全部采用半透明镀膜玻璃，以减轻封闭的感觉。

外立面采用滑窗，可向内滑移，灰色调的拼图使内墙像幅巨大的壁画。

玻璃外立面外的建筑覆盖层，是绝缘外壳，避免出现热桥问题。

在内部种植了一棵大树，使其与浅色拼图区域相交融。除了大开口的外墙，在较短的一边，南北双向布置通向地面的阳台，汽车坡道的上方也是一个不错的公共空间。

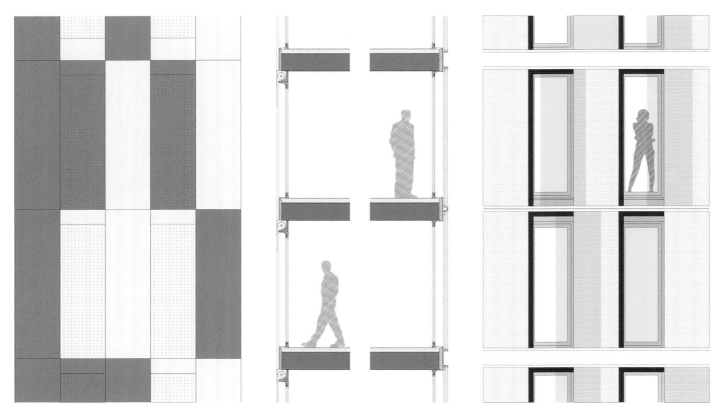

SISTEMA CONSTRUCTIVO

1. FACHADA INTERIOR

1.1. FACHADA TIPO COTETERM: A.PINTURA MINERAL AL SILICATO B.TEXSA COTETERM 2 CAPAS CON MALLA FIBRA DE VIDRIO C.AISLAMIENTO DE POREXPAN e: 7cm. D.IMPERMEABILIZACION ELASTICA E.ENFOSCADO Y MAESTREADO, e: 1.5cm. F.1/2 ASTA LADRILLO PERFORADO G.GUARNECIDO Y ENLUCIDO DE YESO, e: 1.5cm. 1.2. PERSIANA ENROLLABLE ALUMINIO 1.3. CARPINTERIA EXTERIOR DE ALUMINIO LACADO CON ROTURA DE PUENTE TERMICO Y VIDRIO 4/12/5 1.4. REMATE DE ALUMINIO LACADO, INCLUSO FRENTE DE FORJADOS SUP. E INF. PREVIO AISLAMIENTO e: 4cm.

2. FACHADA EXTERIOR

2.1. VIDRIO EXTERIOR LAMINARA 6+6 CON BUTIRAL TRASLUCIDO 2.2. PERFILERIA U ACERO GALVANIZADO LACADO, SUJECION DE VIDRIO EN FACHADA 2.3. PERFIL U ACERO GALVANIZADO LACADO 300.100.6 2.4. CONTRAVENTANA CORREDERA: VIDRIO LAMINAR 6+6 CON BUTIRAL TRASLUCIDO 2.5. CARPINTERIA EXTERIOR DE ALUMINIO LACADO CON ROTURA DE PUENTE TERMICO Y VIDRIO 4/12/5 2.6. PREMARCO DE ALUMINIO LACADO, INCLUSO REMATES DE FRENTE DE FORJADOS SUP. E INF. 2.7. FACHADA TIPO COTETERM: A.PINTURA MINERAL AL SILICATO B.TEXSA COTETERM 2 CAPAS CON MALLA FIBRA DE VIDRIO C.AISLAMIENTO DE POREXPAN e: 5cm. D.IMPERMEABILIZACION ELASTICA E.ENFOSCADO Y MAESTREADO, e: 1.5cm. F.1/2 ASTA LADRILLO PERFORADO G.GUARNECIDO Y ENLUCIDO DE YESO, e: 1.5cm.

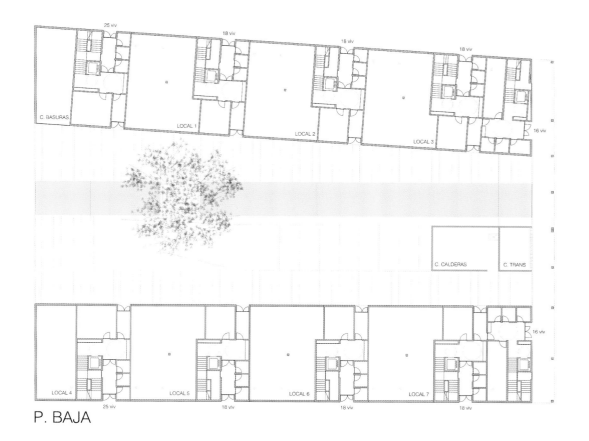

P. BAJA

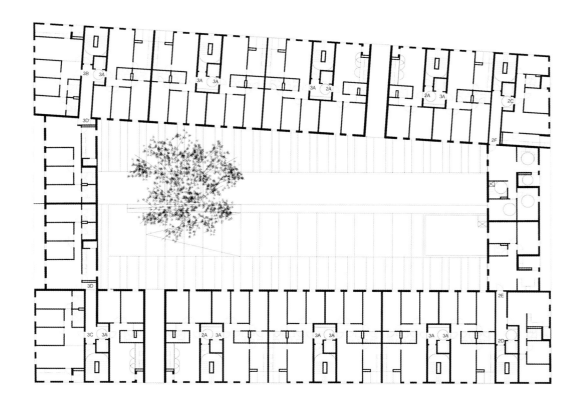

FACADE
立面

TRIANGOLO

TRIANGOLO 公寓

Architect: Sebastian Nagy Architects, s.r.o
Designer: Ivan Matusik, Sebastian Nagy
Client: Stavomex, s.r.o
Location: Spojovacia 30, Nitra, Slovakia
Site Area: 1,000 m²
Status: Completed
Units: 15
Photography: Bobo Boska, Ivan Karlik

设计公司：Sebastian Nagy Architects, s.r.o
设计师：Ivan Matusik, Sebastian Nagy
客户：Stavomex, s.r.o
地点：斯洛伐克尼特拉市 Spojovacia 街 30 号
占地面积：1 000 m²
状态：已建成
户数：15
摄影：Bobo Boska, Ivan Karlik

The original construction cost is €1,300,000 that means: construction cost in China is

¥ 3,759,617

per unit cost is ¥ 250,641

原建造成本为 1 300 000 欧元，
国内建造成本约

¥ 3 759 617 元

国内平均每户造价约 250 641 元。

The apartment house TRIANGOLO was created as a dialog between natural predispositions, town and the construction site locality. The Nitra's upland, Zobor and meander of the river Nitra represented basic inspirational dominants. From typological point of view the building is a three-tract, the corridors of which are lit along the entire height by daylight from terminal vertical skylight. In the architectonic expression dominates mono-materialism, two facades of identical appearance and curved section of the roof with domes. The dispositional and also the constructional principle unroll from bearing ferro-concrete cores, which enabled creation of apartments in various size categories. Individual conception is also indicated by the house symbol located by the entrance to the house.

TRIANGOLO 公寓的建造就像是城镇、建设工地和自然界之间的一场对话。尼特拉高地、Zobor 修道院和缓缓流动的尼特拉河共同演绎着这片土地的优美与灵动。这座建筑是三角形结构，日光透过末端天窗从顶部照亮走廊。为表达出绝对的物质存在，设计师采用两个完全相同的外立面、拱形屋顶的弯曲面。钢筋混凝土支撑核成就了这座尺寸类型丰富的公寓，而房屋入口处的标志则彰显出主人的个性。

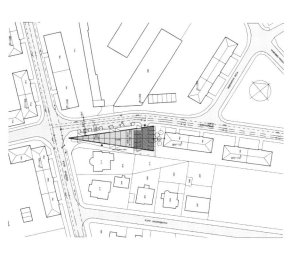

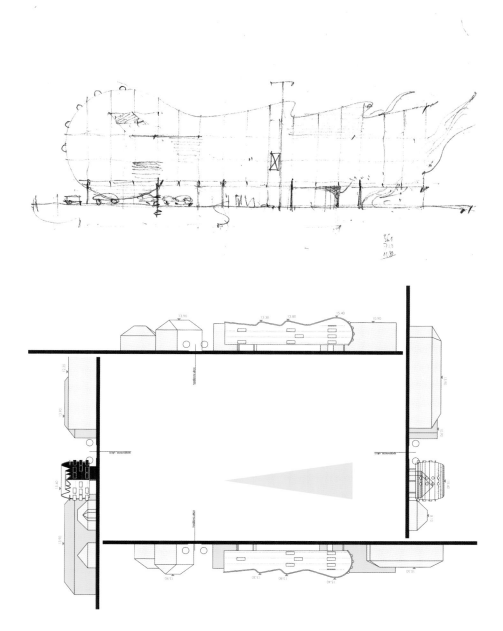

Cost and Creativity 成本与创意

The unique shape not only saves space and materials, but also taking into account the orientation and lighting.

独特的外形不仅节约了空间和材料，还能同时兼顾朝向和采光。

Triangular parcel and the roof of harmonic curve together with dispositional and material concept eventually formed architecture into its final resemblance.

三角形造型、柔美弯曲的屋顶和独具匠心的选料，最终成就了这片土地独特的建筑风格。

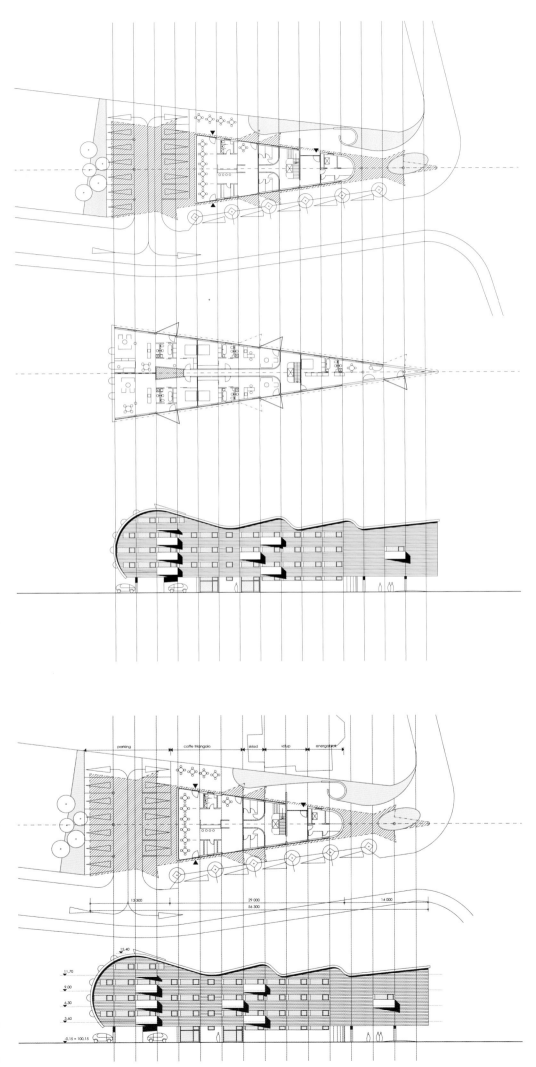

parking   caffe triangolo   sklad   vstup   energoblok

13 300
29 000
56 300
14 000

15.40

11.70
9.00
6.30
3.40

-0.15 = 100,15

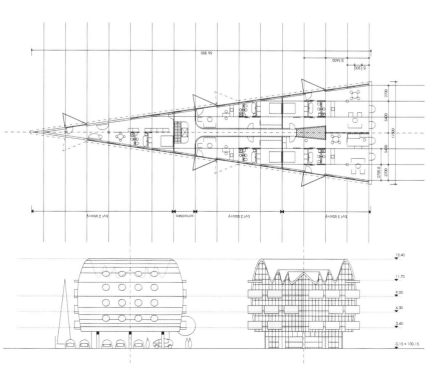

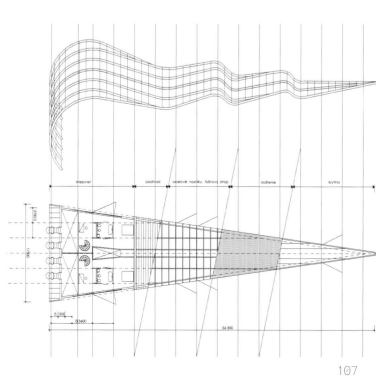

HOUSING IN CHOISY-LE-ROI

CHOISY-LE-ROI 公寓

Architect: Trévelo & Viger-Kohler Achitectes/
        Ubanistes
Client: OPAC Val-de-Marne, Sadev 94
Location: Choisy-le-Roi, France
Site Area: 2,100 m²
Status: Completed
Units: 26
Photography: David Boureau, Guillaume Clément

设计公司：Trévelo & Viger-Kohler Achitectes/
        Ubanistes
客户：OPAC Val-de-Marne, Sadev 94
地点：法国 Choisy-le-Roi
占地面积：2 100 m²
状态：已建成
户数：26
摄影：David Boureau, Guillaume Clément

The original construction cost is €2,180,000
that means: construction cost in China is

¥ 4,179,148

per unit cost is ¥ 160,736

原建造成本为 2 180 000 欧元，
国内建造成本约

¥ 4 179 148 元

国内平均每户造价约 160 736 元。

This project is poised to make a statement about what social housing can be: to provide high quality housing at controlled costs, and promote a strong architectural image in the city.

本案着眼于呈现经济房的新标准：在控制成本内提供高品质的住房，并使其成为城市的一个令人震撼的建筑形象。

PAREMENT BRIQUE CLAIRE
MENUISERIE PVC
ENDUIT
GARDE CORPS GALVA PEINT

PAREMENT BRIQUE SOMBRE

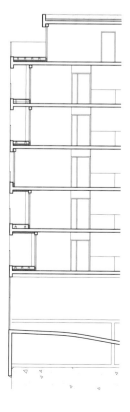

Cost and Creativity 成本与创意

To reduce costs and to control the budget TVK architects worked on the capacity and the simplicity of the building. They also used the brick only on the outside skin and changed to plaster, a cheaper material, for the windows' frames and the terraces' façades.

为了降低成本和控制预算，TVK 的设计师注重建筑容量和简化施工。例如外层表皮只使用砖材，窗框和露台的外墙则用了更低价的材料——石膏。

The choice of a principal, continuous material for the group of façades expresses a sense of unity and reinforces the geometry of the block.

选材的原则以连续型的材料为准，让立面呈现出强烈的统一感，加强了建筑的几何形状。

The 26 units are distributed around a central core stairwell in order to take advantage of the site's angled position. The service rooms are grouped together at the centre of the building, and the living rooms benefit from a double exposition.

The views of the Seine and the surrounding landscape is thus widely visible. The load-bearing façade allows the elimination of bearing walls in the apartments in order to leave open new possibilities over time.

While the apartments are based on a rational principle of organization, the arrangement of the balconies is what gives its unique character to each unit. Each apartment benefits from a private outdoor space: gardens for the ground floor, balconies for the middle floors, and large terraces for the top floors.

The exterior is enriched by the diversity of balconies that create a random visual play on the façade. The utilization of bricks as a facing adds further complexity to this imperfect site, and gives one the perception of a façade that is constantly changing.

根据场地的角度优势，公寓的 26 间房都围绕着中心的楼梯分布。服务室也设置在公寓的中心，客厅则能获得双面的开阔视野。

公寓位置视野开阔，坐拥塞纳河景及其周围的风光。因其出色的承重性能，立面取代了公寓内的承重墙，以便日后重新创造或分配空间。

虽然公寓是在合理的建筑原则上建造，但阳台的设置却赋予了公寓房间最独特的个性。每间房都享有私人的户外空间：首层的房间有花园，中层的有阳台，高层的有大露台。

外观多样化的阳台在立面上创造出随机的视觉效果。设计采用砖石作为覆饰进一步弥补了基址的不足，并给人一种立面在不断变化的感觉。

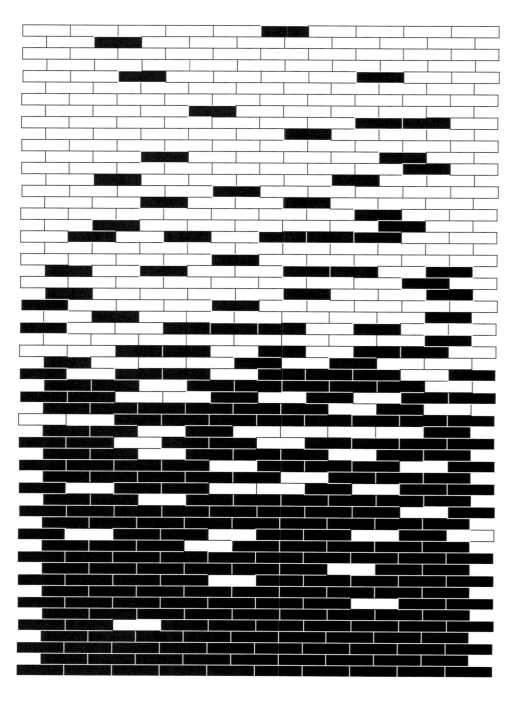

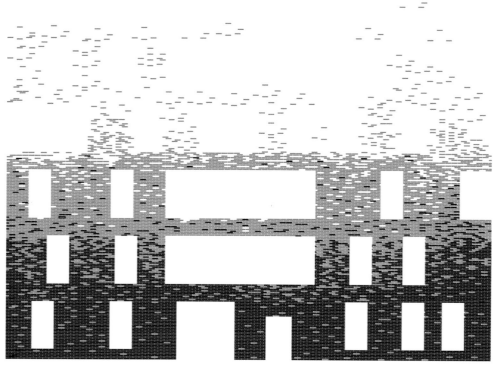

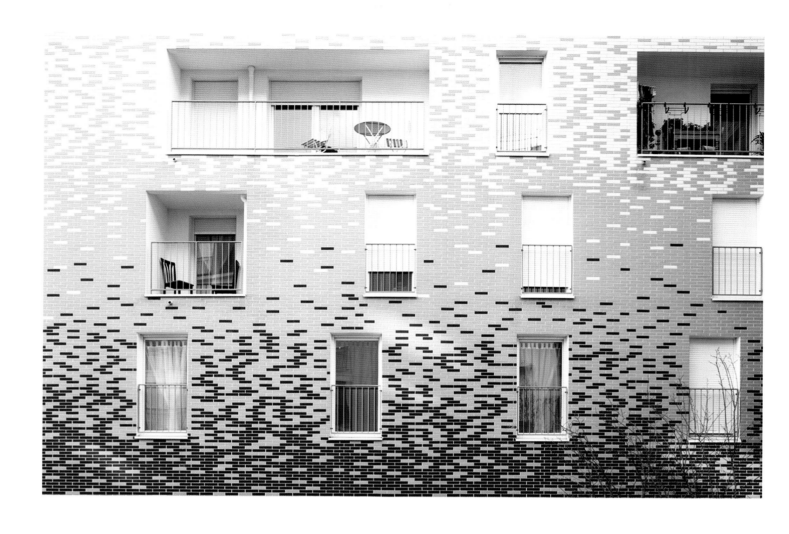

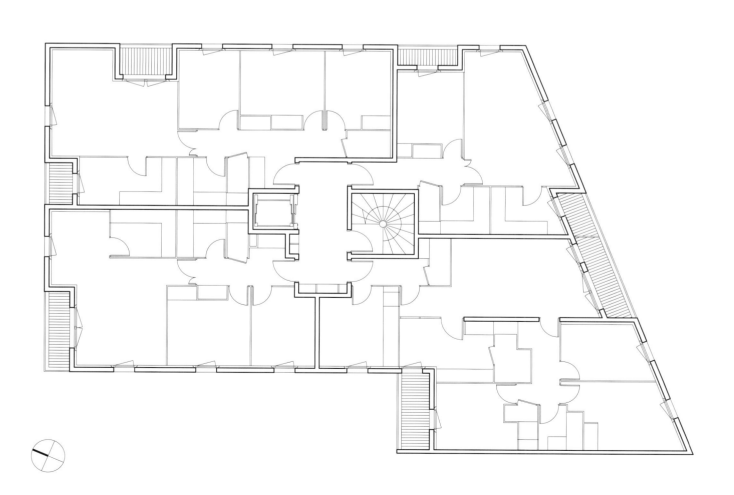

LACE
APARTMENTS

蕾丝公寓

Architect: OFIS Arhitekti
Client: Kraški Zidar d.d.
Location: Nova Gorica, Slovenia
Site Area: 6,500 m²
Status: Completed
Units: 63
Photography: Tomaz Gregoric

设计公司：OFIS Arhitekti
客户：Kraški Zidar d.d.
地点：斯洛文尼亚新戈里察
占地面积：6 500 m²
状态：已建成
户数：63
摄影：Tomaz Gregoric

The original construction cost is €5,200,000 that means: construction cost in China is

¥ 12,235,753

per unit cost is ¥ 194,218

原建造成本为 5 200 000 欧元，
国内建造成本约

¥ 12 235 753 元

国内平均每户造价约 194 218 元。

The location of the apartment block is in the centre of Nova Gorica (population 32.000) – Nova Gorica is situated in the west of Slovenia, adjacent to the Slovene – Italian border. It lies 92 meters above sea level and has very specific climate conditions – it is renowned as the hottest town in Slovenia during summer and also has very strong winds in winter. With its climate, vegetation and way of living the city has a definite Mediterranean character, where external shaded space has high importance. Therefore, the clients demand was to design rich external spaces of varying characters.

本案的公寓坐落于拥有 32 000 人口的新戈里察市中心。新戈里察市位于斯洛文尼亚西部，毗邻斯洛文尼亚与意大利的边境，海拔 92 m。这座城市具有非常特殊的气候条件，在斯洛文尼亚境内以"夏季最热的城市"而闻名，冬季又有强风的侵袭。该市特殊的气候、植物和生活方式具有典型的地中海特征，因此内部的空间就显得尤为重要。本案的客户还要求将公寓的立面设计得富有新意和特色。

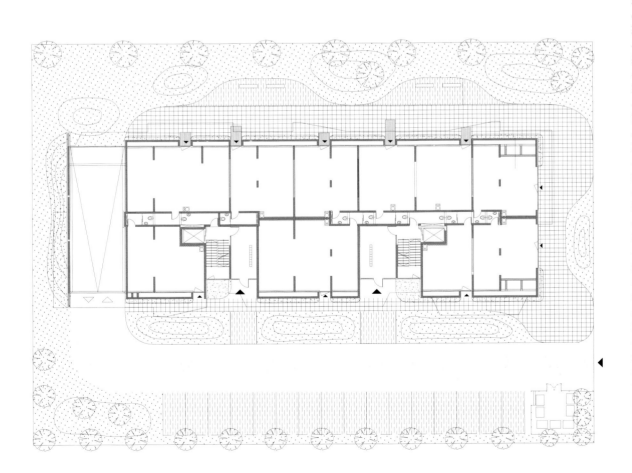

Site Plan

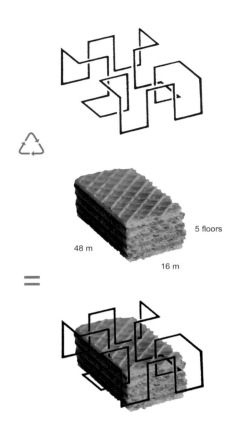

5 floors

48 m

16 m

Cost and Creativity 成本与创意

Materials used are cheap but well made, the buildings' structure is flexible and well organized so that suitable to families in different needs.

公寓选用的材料价格低廉，但品质精良。建筑结构灵活，布局合理，可以适应不同家庭的需求。

Using traditional elements the building reinstates three-dimensional lace which embraces its volume.

传统元素的应用让公寓的立面看起来像是被立体的蕾丝所包围。

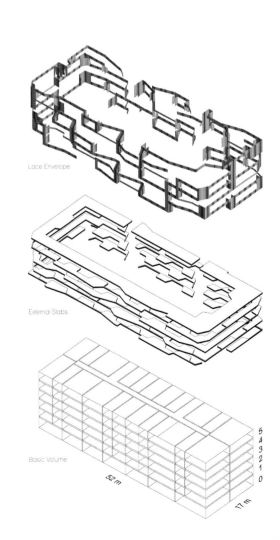

Lace Envelope

External Slabs

Basic Volume

52 m

17 m

The fixed urban plot of the building had to be an orthogonal block 48 m x 16 m x 5 floors. The clients brief was also clear that apartment sizes and typologies were to be simple and repeated. Over this simple structure the second skin of terraces provides each apartment with a different character and a possibility for the buyer to choose the one that responds to his lifestyle. The facade color patterns are taken from typical color elements of the area (typical Gorica valley soil, wine and brick roof-tops) ...nevertheless the locals began to call it "pajamas"(for its similarity to a pajama pattern).

该固定的城市地块将公寓固定为一座长 48 m，宽 16 m，高 5 层的矩形大楼，客户也清楚地要求公寓的大小和类型要简单，并具有一定的可复制性。因此公寓的结构设计简洁，露台作为第二层表皮，让各套房都有所不同，住户可以根据自己的生活方式选择适合的房间。立面的色彩和图案均取自该地的典型色彩元素（戈里察河谷的典型土壤、酒和砖结构天台等），因为其条纹和睡衣很相似，当地人还给公寓起了个别名——"睡衣公寓"。

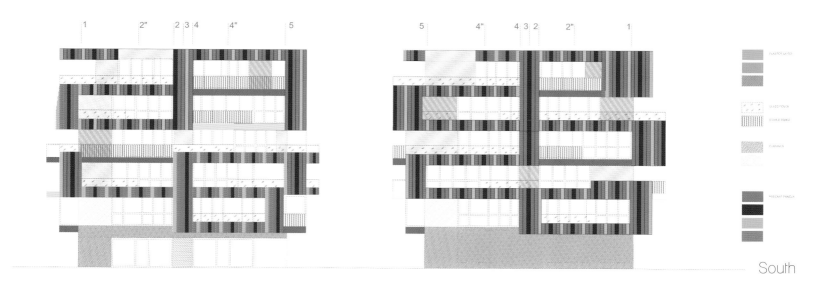

South

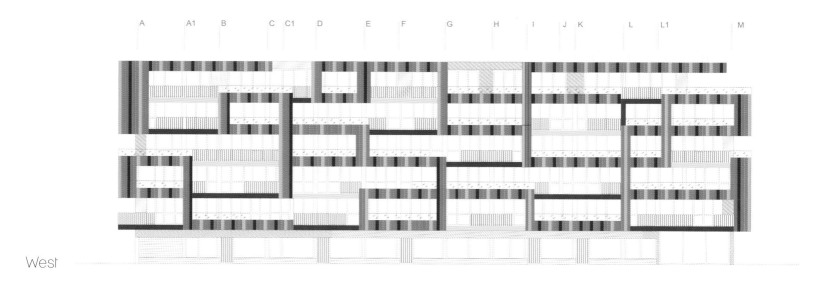

West

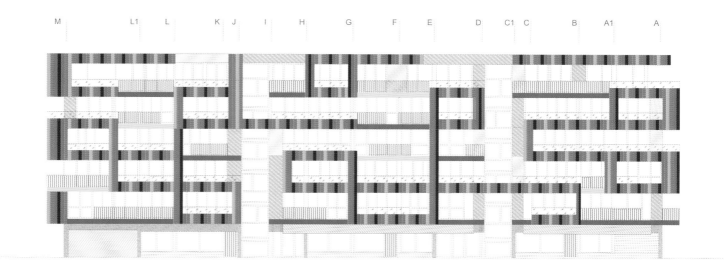

East

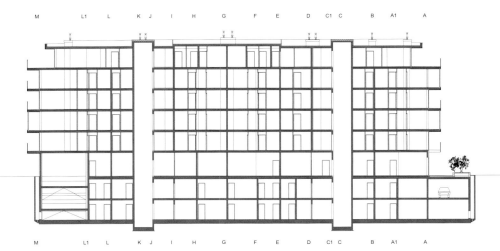

As well as the 5 floors two underground garages were built. In the ground floor, offices and shops/services are located. The combination of façade elements such as, projecting roofs, pergolas, apartment dividing walls, terraces and balconies with loggias function as a constant temperature buffer zone to the main living and sleeping areas and protect against sudden weather changes and strong winds. Additional aluminum shading panels are placed on the outer sides of the winter loggias and balconies. The service and communication spaces are reduced to minimum thus the daylight is provided on the shafts. The basic monthly energy and service costs are very low.

公寓高 5 层，有两个地下车库，首层是办公室、商店和其他服务设施。立面上各种元素的组合，如凸出的屋顶、棚架、分隔墙、配有凉廊的露台和阳台等，都可作为固定的温度缓冲区，为公寓房间的主居住区和卧室阻挡突变的天气和强风。冬季时公寓还会将铝制遮阳板放置在凉廊和阳台的外侧。公寓里服务和交流的空间被缩减到最小，以增加日光的照射。公寓每月的能耗和服务成本也非常低。

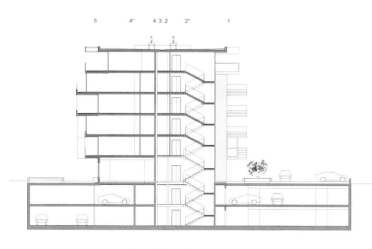

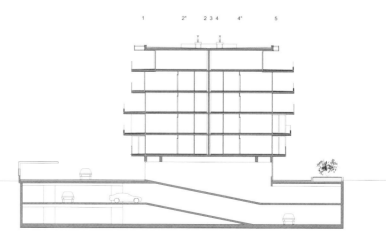

Having studied the different types of external spaces on the existing houses in the area the following types were proposed: balcony and terraces – both opened and covered with roof or pergola, loggias that are closed from the side and fully or partly glazed and fences of different characters – transparent with glass or metal, full and of different heights. Many balconies have integrated cupboards for storing external furniture. By using all of these traditional elements the apartments gain different characters that are more exposed or intimate and that offer opened or partially enclosed views.

经过对该地现有不同类型住房外部的研究，设计师提出了一个方案：阳台和露台为开敞式设计，以顶棚、凉棚或凉廊作为天棚，用玻璃做完全或部分的围合，这样从旁边看就是一个封闭的空间。栏杆也各有特色，以透明的玻璃或金属为材料，或整齐排列，或参差不齐。多数阳台上还配置了橱柜，可以存放户外用具。传统元素的应用赋予了公寓不同的特点，或开放，或私密，为住户提供了或开阔或半封闭的视野。

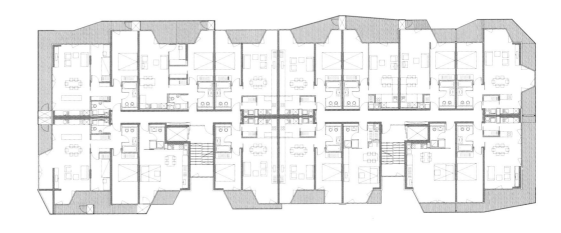

Level 3

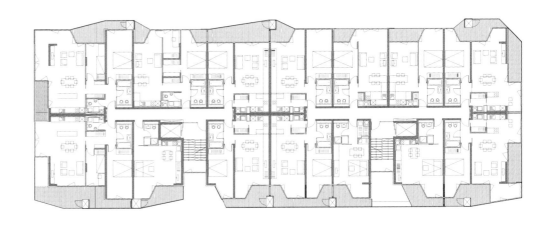

Level 2

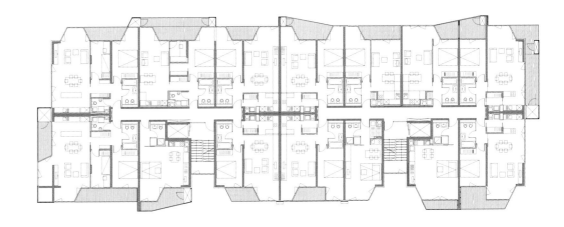

Level 1

BREIAVANNET PARK

BREIAVANNET PARK 住宅楼

Architect: Helen & Hard
Client: Breiavannet Eiendom AS
Location: Stavanger, Norway
Site Area: 540 m²
Units: 14
Photography: Emile Ashley

设计公司：Helen & Hard
客户：Breiavannet Eiendom AS
地点：挪威斯塔万格
占地面积：540 m²
户数：14
摄影：埃米尔·阿什利

The original construction cost is €3,450,793 that means: construction cost in China is

¥ 7,508,243

per unit cost is ¥ 536,303

原建造成本为 3 450 793 欧元，
国内建造成本约

¥ 7 508 243 元

国内平均每户造价约 536 303 元。

The project's location, in the center of Stavanger's historical wooden house district, produced interesting architectural challenges. With the district regulated as a special zone for preservation, the project aimed to knit the language of the local vernacular with that of contemporary architecture.

本案位于斯塔万格的木房子历史风情区的中心，这里有各式各样的木制建筑。因为这是一个受到特殊保护的区域，所以本案设计的最终目的是将当地建筑特色与现代建筑方式巧妙地融合在一起。

Cost and Creativity 成本与创意

Simple materials not only offer for a succinct appearance, but effective in cost savings.

简单的材料既让项目外观简洁，又能有效地节约成本。

The façade in light yellow and white stands out in the surrounding buildings, brightly and eye-catching.

黄白相间的立面在周围住宅楼的包围下格外引人注目。

The building was divided into three volumes creating a rhythm that echos and interacts with the historic houses across the street. The roof forms terminate lower than might be expected, breaking up the volume and creating a relationship with neighboring houses. Wood siding has been used as a façade material while parti-walls and floors are constructed from concrete.

本案被分成三个部分，构成了一个颇具韵律感的建筑形式，与街道对面的历史建筑相互呼应。屋顶比预期建得要低，但与周围的住宅非常协调。外部幕墙采用木材进行装饰，室内墙和地板则使用混凝土建造。

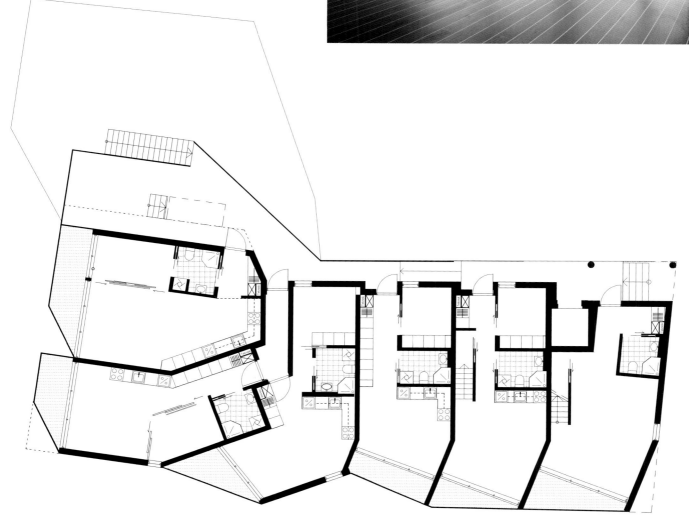

THE ZIELONY GRADUATED STUDENT VILLAGE

ZIELONY 研究生公寓

Architect: Schwartz Besnosoff Architects
& Bar Orian Architects
Location: Haifa, Israel
Site Area: 22,000 m²
Status: Completed
Units: 215

设计公司：Schwartz Besnosoff 建筑事务所、
Bar Orian 建筑事务所
地点：以色列海法市
占地面积：22 000 m²
状态：已建成
户数：215

The original construction cost is $38,000,000
that means: construction cost in China is

¥ 65,284,726

per unit cost is ¥ 303,650

原建造成本为 38 000 000 美元，
国内建造成本约

¥ 65 284 726 元

国内平均每户造价约 303 650 元。

The Zielony Graduate Student Village is located near the main entrance to the Technion, on an area of 22 dunams (22,000 square meters). It includes 215 housing units, ranging from 2 to 4 rooms. The village will also include a community center with preschool classrooms and a multipurpose hall, which will serve as a social hub for the village residents as well as graduate students who live off-campus.

Zielony 研究生村坐落在以色列科技学院的出入口附近一片约 22 000 m² 的区域。这里有 215 套房屋，每套 2 至 4 个房间，村内有公共活动中心，设有幼儿园教室和多功能厅，它不但是村内居民的社交中心，也为居住在村外的研究生提供服务。

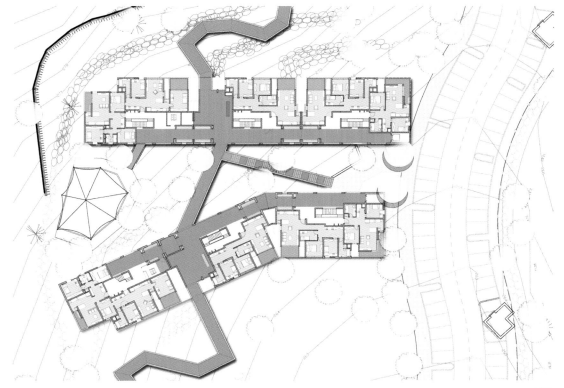

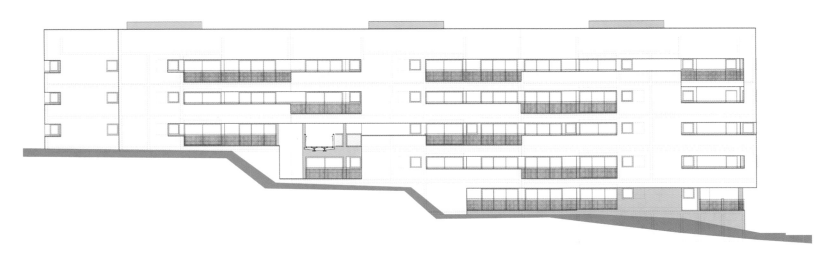

North Elevation

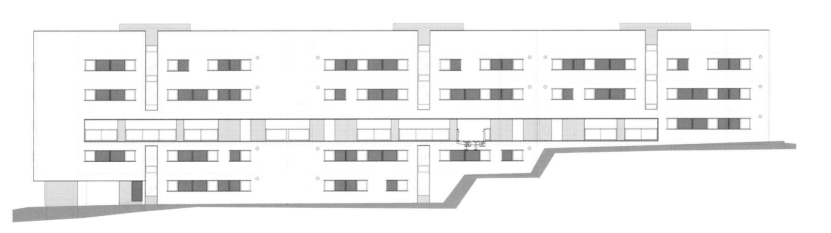

South Elevation

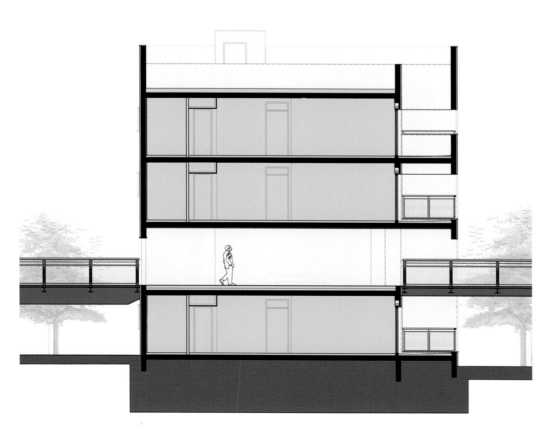

Cost and Creativity 成本与创意

We have reduced the building costs by leaving the slopes around the buildings as they were natural. We preserved the landscape do not used expensive bearing walls. The general building's movement system also considered very few lifts and a cheap maintenance because of it.

在本案中，成本控制的方法包括保留建筑周围的天然斜坡、保留周围的景观、不使用昂贵的承重墙等。建筑的运输系统也少用电梯，并且建筑本身也不需要太多的维护。

The informal construction conveys a sense of freedom, characterized by kibbutz-like buildings and lush green surroundings.

随意的结构传达出一种自由感，kibbutz 式的建筑和郁郁葱葱的绿色环境是其一大特征。

The project design is based on the principles of landscape-sensitive and ecological planning. This is reflected in minimal damage to the landscape, as little development work as possible, preservation of the existing natural environment, and the creation of a silhouette that blends into the natural surroundings.

Pedestrian traffic within the village is accommodated by 7 bridges, built perpendicular to the topography on pillars at the different levels, which lead to a public boardwalk where they join to become a single unit. The boardwalk is the "main boulevard" in the public space of the village. It is the heart of the project – a central meeting place where children can move freely and safely within the village. The definition of the boardwalk as a public space enabled significant savings in the number of elevators and created a unique high-quality and significant public, social space for this community.

The project was developed around the concept of dialogue – with nature, with the neighbors, with the community and with the economy.

本案的设计基于环境敏感和生态规划原则，尽可能减少对景观的破坏，保护现存的自然环境，创立一种融入自然的模式。

村内的行人交通由 7 座桥来解决，它们架设在不同高度的台柱上，通向公共木板道和独立的单元。木板道是村里公共空间的重要"林荫大道"，它是项目的重点，也是聚会中心，孩子们可以在村内安全自由地跑动，木板道作为公共空间有效地省去了不少楼梯，为社区创建了独一无二的高质量、有效的公共社会空间。

项目的发展围绕着对话概念，包括与自然、与邻居、与社区、与经济的对话。

Climate – The building facades are built in a north–south direction.

Ecology – The planning incorporated maximum consideration for the natural green environment.

Minimal incursion on the landscape – The typology of housing clusters built perpendicular to the topography provides maximum distance between them without intervening in the green areas between the housing clusters.

Flexibility – The project site requires flexible construction that can accommodate varying functional, topographical and landscape constraints without the need for essential changes. Along the same lines, the units are planned for flexibility of size (rooms that can be incorporated in different unit configurations).

"Absorbent" buildings – The basic typology of construction is founded on the goal of absorbing the environment – uncultivated surrounding greenery, yards and lanes. The concept of absorption is expressed in enlargement of the building envelope and the opening of "holes" in its volume.

Connection to the soil – The construction plans are based on cautious contact with the soil – the project minimizes the use of supportive walls in order to ensure maximum preservation of nature.

Architecture as a backdrop – The silhouette of the project integrates harmoniously with the natural surroundings, without significant changes to the existing landscape.

Neighborliness – The project creates a living space in a neighborhood-like community atmosphere, with common areas and meeting places. At the same time, each unit enjoys maximum privacy and a view of green landscape.

Locality and integration in the fabric of the Technion – The basic typology resembles that of the existing dormitories at the Technion, with the difference that the buildings are constructed perpendicular to the topography and in clusters, creating new ecological, community and cultural elements.

气候——建筑外立面都建在南北方向上。

生态——建筑的规划也将最大范围的天然绿色环境考虑在内。

对环境的最小破坏——公寓虽然是垂直型的集群模式，但每栋楼之间都拉开了最大间距，以避免对大楼之间的绿化带造成干扰。

灵活性——项目地块要求建筑结构灵活，以适应不同的功能变化和地形景观条件，而不必改变原貌。根据这条原则，房屋的形状也设计得各式各样，体现出建筑的特色，传递出自由自在的感觉，展现出苍翠繁茂的绿色环境。

"吸收型"建筑——建筑的结构是以"吸收环境"为目的，如周边未开发的绿化区、庭院和小巷等。"吸收"的概念主要体现在建筑立面的扩大和体块的"大开口"上。

与土地相接——建设计划考虑到公寓与土地的相接，以尽量减少支撑墙的使用，最大限度地保留自然环境。

作为背景——公寓的轮廓与周围的自然环境和谐相融，不会对原有环境造成太大改变。

社区——公寓打造出了一个具有社区氛围的居住空间，为学生提供共享区和集会区；各个房间在确保居住者的隐私得到保护的同时，又能让居住者享受到户外的绿色美景。

地形和形态的整合——公寓和大学原有公寓的结构相似，为了有所区分，新公寓被打造成垂直的形态，并以集群的形式创造出新的生态、社区和文化元素。

KIRAL
APPARTMENTS

KIRAL 公寓

Architect: Arqmov Workshop
Location: Mexico City, Mexico
Built Area: 2,000 m²
Status: Completed
Units: 17
Photography: Rafael Gamo

设计公司：Arqmov 工作室
地点：墨西哥墨西哥城
建筑面积：2 000 m²
状态：已建成
户数：17
摄影：Rafael Gamo

Kiral (chirality = is the geometric property of a rigid object of being non-superposable on its mirror image; such an object has no symmetry elements of the second kind) is the name that identifies and transmits personality to the residential building designed by ARQMOV Workshop. It is located in Colonia Juárez, Mexico City, a few blocks away from the traditional Reforma Boulevard (Paseo de la Reforma), a cultural and touristic corridor in the most important and populated city of Mexico.

本案由 ARQMOV 工作室负责设计，Kiral 的意思是在普通公寓楼上彰显并传达自己的个性。本案坐落于墨西哥城的 Colonia Juárez，穿过几个街区就可以到达墨西哥城中最重要且人口最密集的文化和旅游长廊——Reforma 大道。

Cost and Creativity 成本与创意

It was a regular construction with nice finishes; we knew since the beginning that we had to invest a little bit more in finishes and particularly in the facade design. We had an original budget that we had to maintain in order to give our investors good profits.

本项目运用常规的施工过程，细节严谨；但是设计公司在一开始时就知道，为了完成细节部分和外观设计，必须自己进行一部分的投资。为了给投资者带来可观的利润，设计公司必须在原来的预算中完成整个项目。

The dynamic façade moves as the city itself is moving.

动感的波浪立面，仿佛让城市也随之舞动。

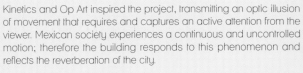

Kinetics and Op Art inspired the project, transmitting an optic illusion of movement that requires and captures an active attention from the viewer. Mexican society experiences a continuous and uncontrolled motion; therefore the building responds to this phenomenon and reflects the reverberation of the city.

By the geometric interplay of forms, a dynamic course is achieved, focusing on the main façade as a significant backdrop for the viewer. The horizontal and vertical lines of the balconies create a perfectly modulated grid. This grid is then perceived in a three-dimensional plane where overlapping spherical shapes are once subtracted, then extruded, resulting in a façade that moves. It moves depending from which angle the observer approaches the building; it moves responding to a play of light and shadows throughout the day; it moves, as the city itself is moving.

动力学和视幻艺术是本案的灵感来源，波浪起伏的立面创造出视线错觉，极为引人注目。墨西哥社会曾经历了接连不断的纷乱动荡，公寓也以此立面反映了此现象及其对城市的影响。

设计师通过几何形态的相互作用，使立面产生了动态的效果，让路人无不驻足欣赏。阳台上水平和垂直的线条完美地组成了可调整的网格。在三维模型中以重叠的球形做标准，让网格或凹进，或凸出，从而形成了动感十足的立面。动态的角度会随着不同的观察角度发生变化，从而产生出变化多样的光影效果，就好像城市本身也在舞动一样。

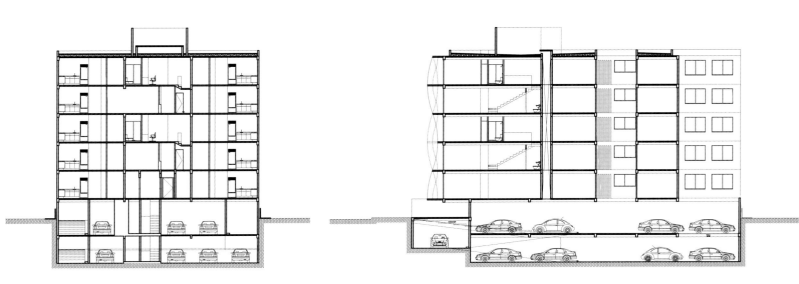

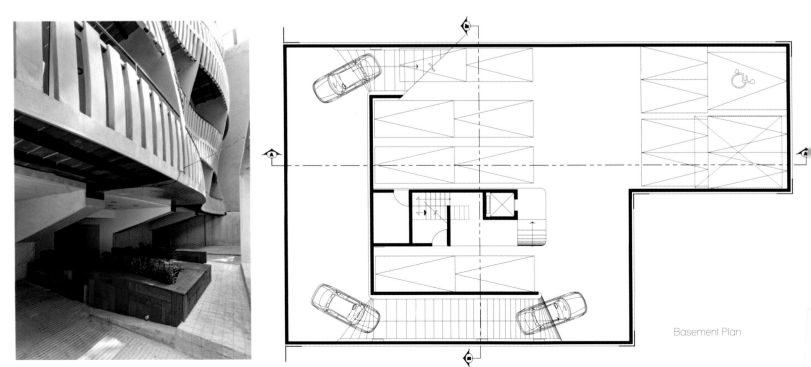

Basement Plan

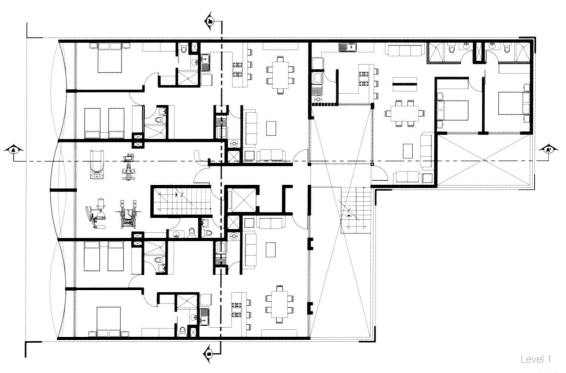

Level 1

The 17 apartments are arranged around a common courtyard, which allows natural light and cross-ventilation. A gym, a roof garden complement a thorough design based on the logical principle of organization. There are one and two-storey apartments with balconies that give the building a sense of community and harmonious coexistence. Every apartment benefits from a private indoor and outdoor space, sharing main services, and maintaining privacy at the same time.

公寓共有 17 套房间，围绕着一个共同的庭院，这样有利于自然采光和交叉通风。鉴于规划时的逻辑原则，本案增设了健身房和屋顶花园，让公寓的功能更为完善。公寓中有部分单层或双层的套房配有阳台，加强住户间的交流，让住户之间能和谐相处。每间套房都享有私人的室内和室外空间，功能服务区虽然是公共的，但也重视保护隐私。

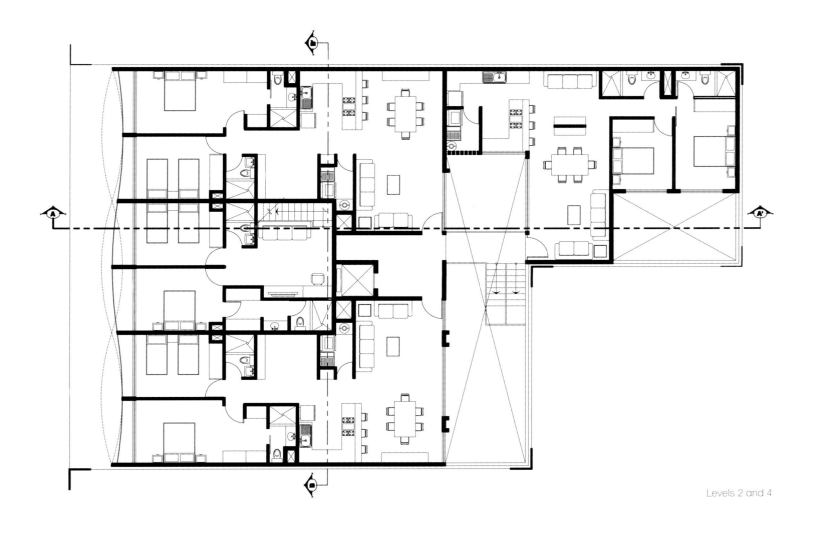

Levels 2 and 4

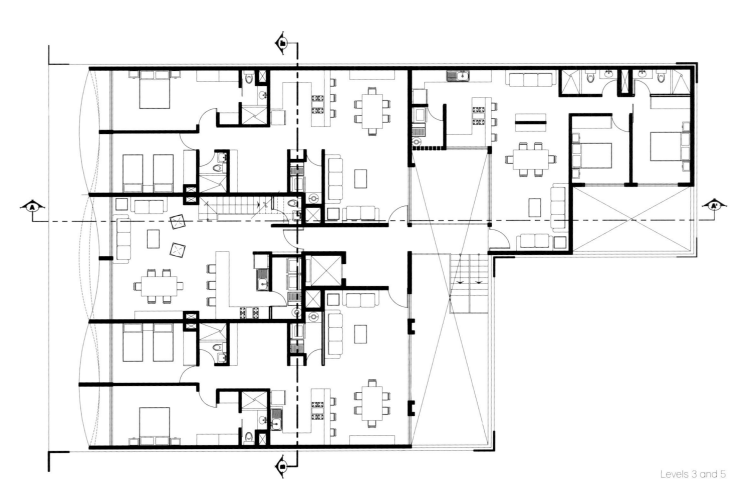

Levels 3 and 5

72 COLLECTIVE HOUSING UNITS

72 房综合型住宅

Architect: LAN Architecture
Client: Ataraxia, Saemcib
Location: Bègles, France
Built Area: 6,500 m²
Status: Completed in 2012
Units: 72

设计公司：LAN 建筑事务所
客户：Ataraxia, Saemcib
地点：法国贝格勒市
建筑面积：6 500 m²
状态：2012 年建成
户数：72

The original construction cost is €7,500,000 that means: construction cost in China is

¥ 14,377,803

per unit cost is ¥ 199,692

原建造成本为 7 500 000 欧元，国内建造成本约

¥14 377 803 元

国内平均每户造价约 199 692 元。

The project's richness and major interest lie in the possibility of inventing an urban lifestyle set in a highly experimental framework enabling the affirmation of new ecological and contemporary architectures. The diversity of architectural propositions and communal and private spaces had to ensure and enhance this specificity.

本案的丰富性和关注点在于以实验性的结构来创造城市生活，为社会带来新的生态型现代建筑。建筑的结构呈多样性，丰富的公共和私人空间让此特性更为明显。

Column-Slab Structure

Prefabricated Panels
without Perforated Metal Sheet

Prefabricated Panels
with Perforated Metal Sheet

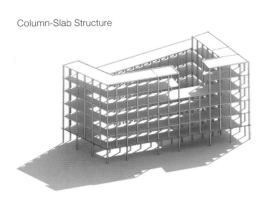
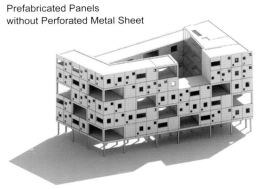
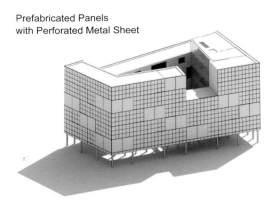

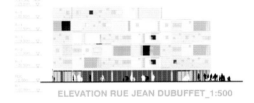

ELEVATION RUE DES PRUNIERS_1:500

ELEVATION RUE JEAN DUBUFFET_1:500

ELEVATION RUE DES MURIERS_1:500

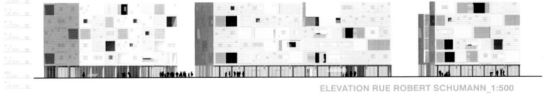

ELEVATION RUE ROBERT SCHUMANN_1:500

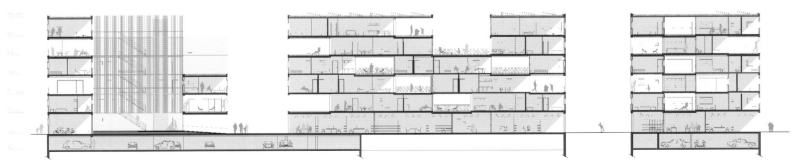

Section CC (North-South) 1:500

C

T4 80m² T2 48m² T2 48m² T2 50m²
T3 69m² T2 51m²
T3 72m² T3 69m²

BLOCK D1
ACTIVITES 365m² LV 30m²
T3 69m²
T2 49m²
T3 61m²
BLOCK D1

T3 88m²
T5 99m² T3 71m²
T4 88m² T3 68m²
T3 60m²

BLOCK D2
LV 32m²
T2 51m²
T2 51m²
T2 51m²
BLOCK D2
ACTIVITES 335m² LV 12m²
PARKING

T3 69m²
T3 60m²
T3 69m² T2 45m² T2 45m² T3 68m²

BLOCK D3
LV 32m²
PARKING
ACTIVITES 410m²

C

☐	**T2**
▨	**T3**
▨	**T4**

🕐 Plan R+2 (+8.50 NGF) 1:500

☐	**T2**
▨	**T3**
▨	**T4**

🕐 Plan RDC (+5.00 NGF) 1:500

Cost and Creativity 成本与创意

Selection of the most suitable materials and consideration of local conditions, and in the support of the flexible concepts and advanced technology, a new housing with low-cost, high-quality is perfectly achieved.

结合当地条件选择最合适的用材，在灵活的理念和先进技术的支持下，完美实现了低成本、高品质的新住宅。

A new, ecological and social living space geared to the 21st century.

面向 21 世纪的新型生态生活空间。

The first stage was to "sculpt" the volumes in order to exploit their urban potential and intrinsic spatial qualities. We directed our research towards a hybrid typology combining the house and the apartment. The principle underlying our approach was that of stacking containers, and careful study of habitat modes, climatic conditions and the sun's trajectory throughout the year suggested the way to organise this.

The project's column-slab supporting structure has a system of lightweight façades providing ultra-high performance insulation levels.

The relative narrowness of the buildings dictated a strategic search for compactness. The idea of variable compactness introduced the notion of a housing unit's adaptability to seasons and times of day. All residents have the possibility of using their exterior space as a windbreak, a mini-greenhouse or, conversely, as a means of cooling or ventilating.

The morphology of each unit stems from the wish to develop housing units enabling a variety of uses very simply and with no extra technological input. We are therefore proposing cross-building units with adaptable exterior spaces and at least two different orientations.

第一步是要挖掘建筑的城市潜力和提高内在的空间质量。设计师的研究方向是要建造一座综合型住宅建筑。建筑的规划原则包括堆叠的方式、当地的生活方式、气候条件、一年中太阳运动的轨迹等，作为建设的基本前提。

本案采用板柱结构作为支持构架，具有立面轻质、高效隔热的特点。

建筑相对狭窄，因为采取了紧密型的设计策略。可变的紧密型理念可以让每户房间都能根据季节和一天中不同的时间做出适应和调整。所有的居民都能将外部空间作为防风墙、小型温室，或是作为降温或通风的方式。

房间的形状各不相同，这样可轻易实现多种用途，并不需要过多的技术投入。因此设计师提出建筑交叉房间的概念，以创造适应性能佳的外部空间，并保证建筑至少有两个不同的朝向。

58 HOUSING UNITS

法国 58 房公寓

Architect: LAN Architecture
Client: Nacarat
Location: Boulogne-Billancourt, France
Site Area: 4, 639 m²
Status: Completed
Units: 58
Photography: Julien Lanoo

设计公司：LAN 建筑事务所
客户：Nacarat
地点：法国布洛涅 - 比扬古
占地面积：4 639 m²
状态：已建成
户数：58
摄影：Julien Lanoo

The original construction cost is €7,200,000 that means: construction cost in China is

¥ 13,802,691

per unit cost is ¥ 237,977

原建造成本为 7 200 000 欧元，
国内建造成本约

¥ 13 802 691 元

国内平均每户造价约 237 977 元。

The major challenge for the design of the housing units on plot V was to be able to resolve all the problems through the use of a single architectural gesture: 1. design of a corner building. 2. positioning on a plot closing off a trapezoid form and the main artery through the space. 3. insertion into a plot with a very restrictive morphology and orientation. 4. continual care taken to ensure integration into an architecturally very rich context using a wide range of materials.

As a result, we imagined a dual character architecture, designing a building able to integrate the urban needs of rue Yves Kermen and rue et Emile Zola, as well as those of the landscape provided by the main artery.

第五地块的住宅单元设计主要问题是需要能使用单一的建筑手法解决所有问题：1. 角楼设计。2. 设计一个封闭的梯形开口，主通路直通空间。3. 采用限制性形态和方位性将其加入图纸。4. 选用各种材料，以持续关注整体环境。

因此，设计师设想一种双重特色的建筑，能够满足城市需求和与主干道景观保持连续性。

FAC03 : COUR NORD

FAC04 : ALLEE ROBERT DOISNEAU

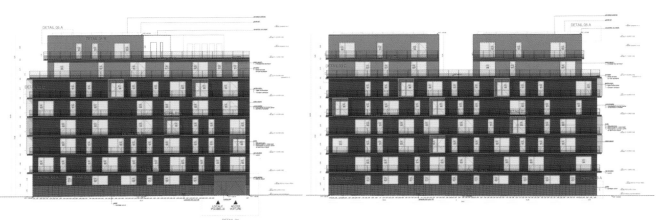

FAC02 : AVENUE EMILE ZOLA

FAC01 : RUE YVES KERMEN

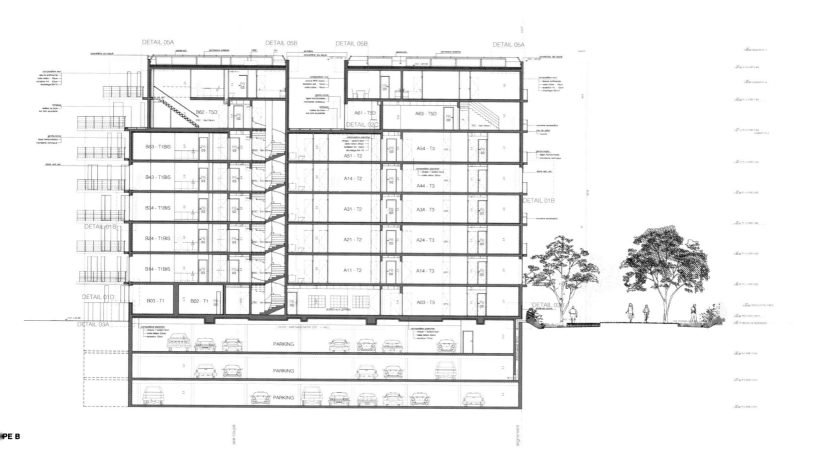

DETAIL 05A · DETAIL 05B · DETAIL 05B · DETAIL 05A

B62 - T5D

DETAIL 02C

A61 - T5D · A63 - T5D

B53 - T1BIS · A51 - T2 · A54 - T3

B43 - T1BIS · A14 - T2 · A44 - T3

DETAIL 01B

B34 - T1BIS · A31 - T2 · A34 - T3

DETAIL 01B

B24 - T1BIS · A21 - T2 · A24 - T3

B14 - T1BIS · A11 - T2 · A14 - T3

DETAIL 01D

B03 - T1 · B02 - T1 · A03 - T3

DETAIL 03A

BOITES AUX LETTRES

DETAIL 03B

PARKING

PARKING

PARKING

PE B

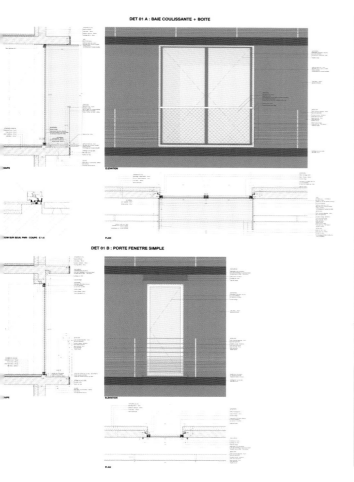

DET 01 A : BAIE COULISSANTE + BOITE

DET 01 B : PORTE FENETRE SIMPLE

Cost and Creativity 成本与创意

To the simplicity of the drawing is added the richness of the material. We have sought for a finish that is clean, shaven and refined but rich enough to reflect light, and communicate with the colours of its environment.

建筑样式简单，材料丰富。设计师追求整洁、光滑和精致的外观，既能反射光线，又能和周围环境的色彩融为一体。

The project proposes a simple and easily readable building giving onto the street.

本案的建筑风格简洁、悦目，在街头形成了一道独特的风景。

The language sought for the elevations on rue Yves Kermen and rue Emile Zola was achieved by designing a corner building, developing a "discreet elegance", creating an event, giving the elevations a certain rhythm and providing an overall impression of lightness.

The south and east elevations are provided with a considerable number of large openings that extend the inside of the housing units out onto the decks running along the side of the building. Wide balconies cantilevering out from the decks interrupt this linear flow.

On rue Kermen and rue Zola, the projecting elements and trims between floor levels give the building the impression of a streamlined block. The building fulfils its role of being a "corner" on the plot and, as a result, clearly stands out from the surrounding buildings.

The language chosen for the courtyard and the main artery through the plot reveals the intention to reverberate and reflect light through the planted areas, favouring a peaceful and serene setting.

在整体设计方面，设计师在表现上力求通过设计一栋角楼，营造低调、优雅的设计项目，使设计正面图呈现出一种节奏，给人以轻盈的整体印象。

南向立面和东向立面设计了不少开口，从住宅单元内部一直伸展延续到楼侧面的平台。从平台上悬伸出的大阳台打断了这一流线。

在该地区，楼层间的项目元素和装饰形成楼房流线型外观。该建筑物实现了作为场地中"角楼"的作用，矗立于周围的建筑物中。

庭院和主干道的设计表现手法是，通过植物区域吸收光线，营造出平和静谧的环境。

KSM APARTMENT BUILDING

KSM 公寓

Architect: Querkraft Architekten zt gmbh
Client: Bauhilfe wohnbau karree st marx projekt gmbh
Location: Vienna, Austria
Site Area: 11,570 m²
Status: Completed
Units: 121
Photography: Lisa Rast, Manfred Seidl

设计公司：Querkraft Architekten zt gmbh
客户：Bauhilfe wohnbau karree st marx projekt gmbh
地点：奥地利维也纳
占地面积：11 570 m²
状态：已建成
户数：121
摄影：Lisa Rast, Manfred Seidl

The original construction cost is €16,500,000 that means: construction cost in China is

¥ 26,589,481

per unit cost is ¥ 219,748

原建造成本为 16 500 000 欧元，国内建造成本约

¥ 26 589 481 元

国内平均每户造价约 219 748 元。

The project is an affordable, flexible, "growing" apartments – especially for young families. Notably in 3rd district there was a limited supply of affordable apartments for young families.

本案是一座经济适用型公寓，灵活度高并且能不断"成长"，特别是在住房供不应求的第三区，非常适合于年轻家庭居住。

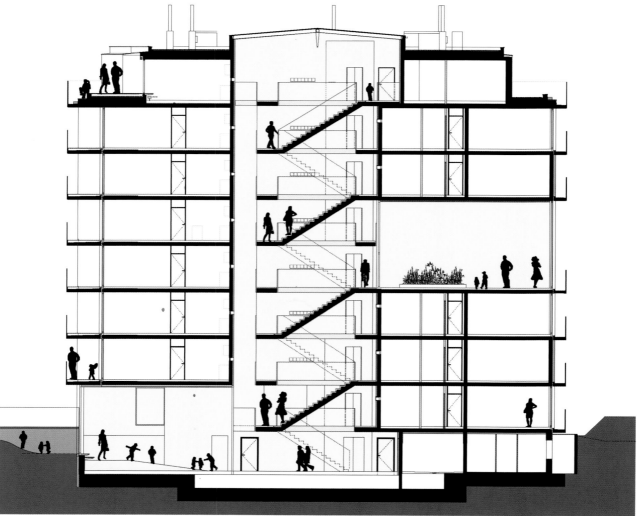

166

The flexible customed elments not only save spaces, but also eliminate the cost of multiple facilities installation.

灵活多变的自定义元素既能节省空间，又能减少安装多种设施的费用。

The homeliness façade offers changeable elements for residents' prefences.

看似朴素的立面给住户提供了众多可变元素，可依据自己的喜好自由设置。

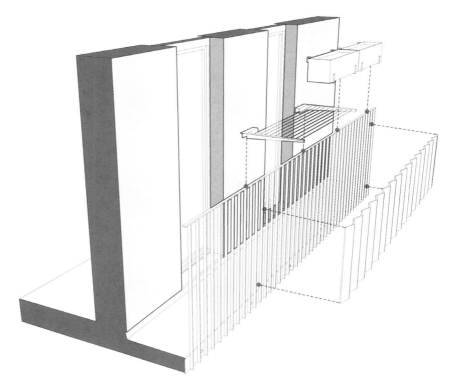

The concept of the windmill makes the double orientation in all apartments possible. The corner apartments are by nature lit and ventilated on two sides. The floorplan orientation means that there are no apartments with an exclusive northern orientation.

Glass panels stretch from floor to ceiling with a disruptive window sill. Living rooms are generously glazed across their entire length. Living spaces are expanded with the addition of perimeter balconies and loggias. Instead of architectonic elements the building budget allows for the residents to receive three elements to design their balconies: flowers pots, a tarpaulin screen and clothes rack. These everyday elements create the first layer of homeliness which will be continued by the residents themselves. Thus the design of the house lies in the hands of the residents and will change continually.

设计采用"风车"的概念，让公寓楼里的每间房都能拥有双重朝向。即使是拐角处的房间也拥有充足的采光和良好的通风。平面布置也经过考虑，避免了房间完全朝北。

玻璃嵌板上嵌大型的落地窗，从地板一直延伸到天花板。客厅的边缘也以玻璃作为墙面，再加上阳台和凉廊，可扩大生活空间。虽然公寓的预算有限，但设计师不是通过建筑元素，而是采用开放的形式，提供了3种元素，让住户设计自己的阳台：花盆、挡雨板和衣架。这些日常的元素不仅可以让公寓看起来较为简朴，还能让用户长期使用并自由发挥。因此，房间的设计权其实是在住户的手中，并可以做众多变化。

In order to maintain the clarity of the town planning basis the private perimeter freespace was removed from the building volume. Loggias structure the body from the outside, a large airy atrium creates a lively communication space on the inside.

Double heighted winter-gardens within the building deliver not only more natural light into the atrium, but also an extension semi-private interaction zone. The main theme of the project was to make the internal atrium a great communications zone. The recesses in the roof, the opening of the green salons and the generous open-plan ground floor, together with the kitchen glazing create a tension filled ever changing light-show every evening. Generous communal functions open the atrium in the ground floor. The transition from the public street area to the private apartment occurs in the atrium itself. The foyer and atrium become a social meeting point for the residents.

The added extras in the green concept for KSM Apartment Building is complemented with an attractive garden concept for the roof top.

为了维护城市规划的原则，公寓楼中并没有设置外部的自由空间，而是从外部用凉廊作为主体结构。大型的中庭创造了一个轻松的内部交流空间。

公寓内两倍高的冬季花园不仅能保证更充足的自然光线射入中庭，还能作为半私人的互动空间。本案的首要目标是建造一个方便住户交流的内部中庭。凹进的屋顶、开放的绿色沙龙、宽敞的首层空间和以玻璃为墙的厨房，都旨在创造一个光影变幻、让人身心放松的空间。首层中庭的主要功能是提供宽敞的交流空间，也是从公共街道过渡到私人公寓的场所。门厅和中庭因此也作为居民的社交空间。

为了宣扬绿色环保的理念，本案的屋顶上还增设了一个漂亮的花园屋顶。

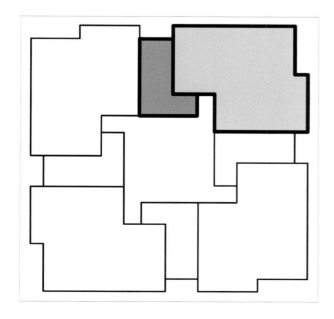

112 FLATS EXPO

世博会 112 公寓

Architect: Basilio Tobías Pintre, Achitect
Client: UTE 112 Viviendas Expo
Location: Avenida de Ranillas, Zaragoza, Spain
Built Area: 18,085 m²
Status: Completed
Units: 112
Photography: Pedro Pegenaute

设计公司：Basilio Tobías Pintre, Achitect
客户：UTE 112 Viviendas Expo
地点：西班牙萨拉戈萨 Avenida de Ranillas
建筑面积：18 085 m²
状态：已建成
户数：112
摄影：Pedro Pegenaute

The original construction cost is €12,781,524 that means: construction cost in China is

¥ 28,126,978

per unit cost is ¥ 251,134

原建造成本为 12 781 524 欧元，国内建造成本约

¥ 28 126 978 元

国内平均每户造价约 251 134 元。

The building's position in the plot established by the urban plan allows the creation of a wide open space, partially landscaped, that will make possible both a grateful use during the Expo and after it as a building of private flats. These open spaces adjust the relationship of the building with the adjoining streets.

公寓的位置由城市规划方案确定，建立在开阔的空地上，重视环境景观的建设，目的是既能为世博会服务，结束后又能作为私人公寓使用。开阔的空间也有助于缓和公寓和毗邻街道间的关系。

Cost and Creativity 成本与创意

The structural and construction system is quite strict. Moreover, the developer was at the same time the builder allowing working together during the construction process in order to adjust costs.

本项目具有严谨的结构和建筑系统。此外，在施工过程中，开发商和施工队都被要求同时在场，以方便随时沟通调节成本。

The building has an L shape and double façade for the relationship between the overall layout and the constructive system.

根据总体布局和建筑系统，公寓的布局成"L"形，并拥有双重立面。

The building is composed of two wings that hold the two main directrix of the plot. That makes the L shape plan face south, towards the wide spaces that limit Ranillas Avenue, counting on views of the river and the north façade of the city core. The building contributes to finish an unfinished part of the city.

The starting point has been to make a general shape nine floors high that emphasize the ensemble's horizontality, according to the crossroad in which the building is constructed. Within this shape a work on the profile has been done making two cuts in the last four floors both in the vertex of the plan and in the east end. The central cut articulates the joining between the wings while the east end's lower height allows for a graduating of the relationship with the street and the neighbour Health Centre.

本案的公寓以场地的两条准线为基准，两座翼楼分别向两侧展开。这样的设计便于公寓的"L"形部分坐北朝南，面向 Ranillas 大道旁的开阔空间，坐享悦目河景和城市中心北部的繁华景象。公寓也是城市发展计划的一部分。

公寓边是个十字路口，设计的出发点是要强调这栋 9 层高公寓的横向组合形态，并以此为基础确定公寓的外观构造，对公寓的下 4 层与上层做了明显的区分。中央的准线将两座翼楼联系起来，东座略低，有助于拉近公寓与街道、社区医疗中心的距离。

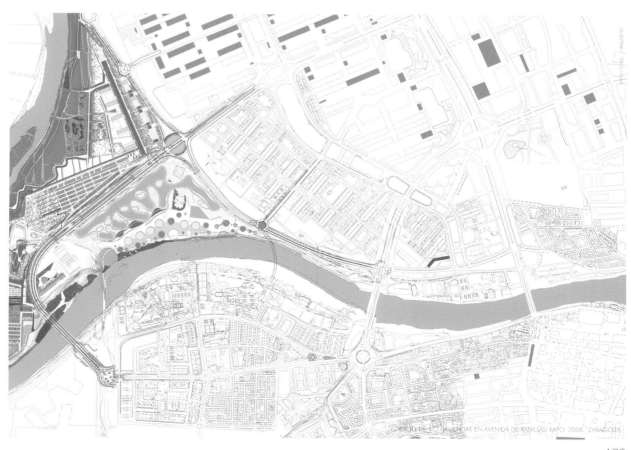

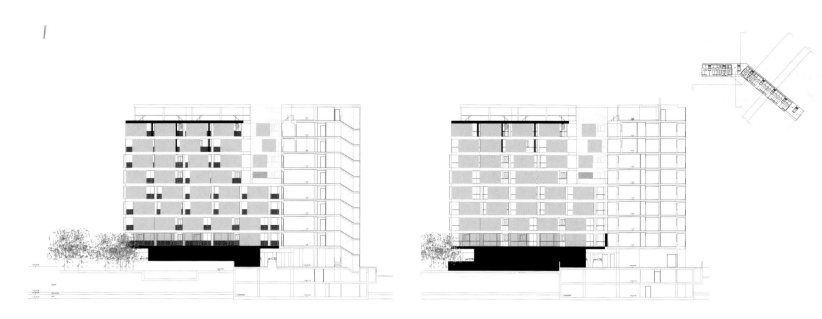

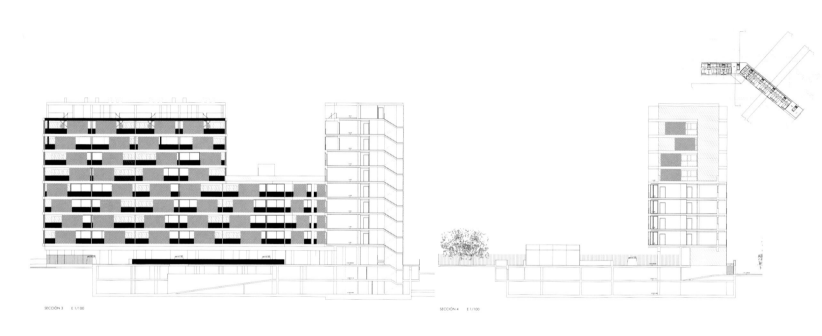

SECCIÓN 3   E 1/100

SECCIÓN 4   E 1/100

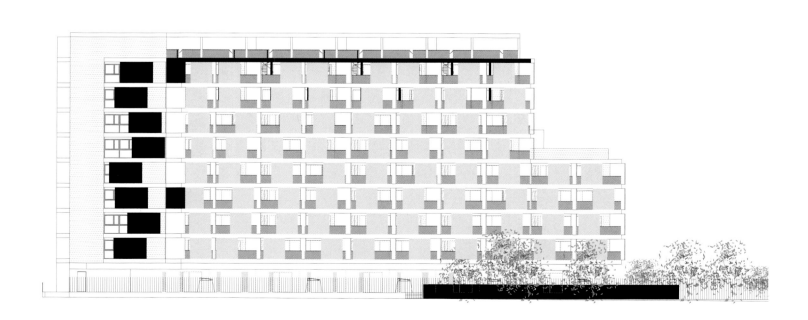

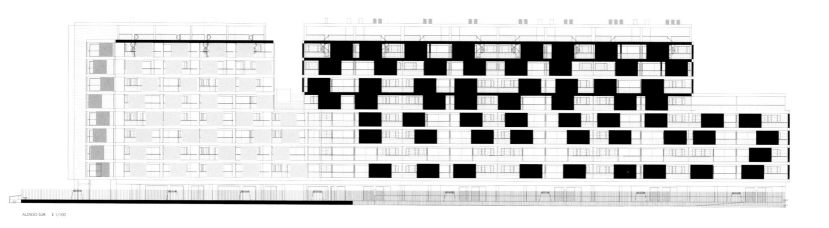

ALZADO SUR   E 1/100

The shaping of the building comes from the relationship between the overall layout and the constructive system that defines the encircling made up of walls clad with long grey metallic ceramics panels that define a background on which longitudinal cantilever slabs have been made, provided with sliding aluminium louvres panels that make up double façades, behind which glass rails have been placed.

公寓的立面造型由总体布局和建筑系统决定。表层是镀层外墙，以灰色的长形金属陶瓷面板作为底板，在纵向设置悬臂板，如此一来就可在悬臂上安装滑动铝制百叶窗作为公寓的第二立面，后面有玻璃护栏。

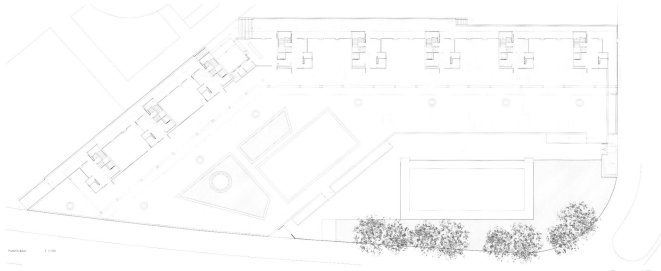

PLANTA BAJA   E 1/100

Ground Floor

168 SOCIAL HOUSING IN MADRID

马德里 168 经济住房

Architect: Coco Arquitectos
Client: Madrid Municipal Housing Agency
Location: Madrid, Spain
Site Area: 4,500 m²
Status: Completed
Units: 168
Photography: Miguel de Guzmán, Ignacio Izquierdo, Coco Arquitectos

设计公司 : Coco Arquitectos
客户 : 马德里市住房局
地点 : 西班牙马德里
占地面积 : 4 500 m²
状态 : 已建成
户数 : 168
摄影 : Miguel de Guzmán、 Ignacio Izquierdo、 Coco Arquitectos

The original construction cost is €9,200,000 that means: construction cost in China is

¥ 20,245,489

per unit cost is ¥ 120,509

原建造成本为 9 200 000 欧元，
国内建造成本约

¥ 20 245 489 元

国内平均每户造价约 120 509 元。

This project is the result of the winning 1st prize in a competition promoted by the Municipal Housing Agency in Madrid, to build 168 social dwelling. The strong sloped plot was placed in the very edge of the city.

Outlying contexts surrounded by city border highways, fields, malls... require consistent answers. These are places where any action demands the same audacity, for example, a leaning building. The construction expresses with the language of kinetic, fitting its shape parallel to the slope while the ground floor retrieves the level of the plot defining streets.

该项目位于城市边缘的一块极为倾斜的坡地上，是马德里市住房局举办的 168 套经济住房竞赛中的一等奖作品。

公寓的周边有城市边境高架路、田地、购物中心等，因此需要与之协调的设计方案。这样环境也需要设计有一些大胆尝试，例如一座倾斜的房屋。公寓的建筑语言生动活泼，形体与斜坡相平行协调，同时地面层轮廓重新定义了街道。

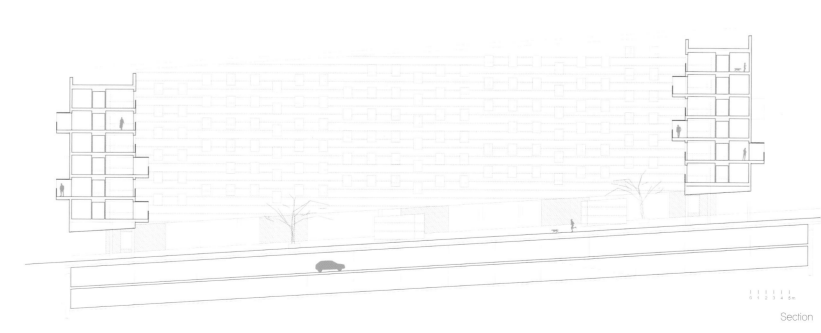

Section

Cost and Creativity 成本与创意

If you have clear ideas and you draw twice as much every plan, it is easy to avoid increasing budgets during the construction.

设计师认为，如果有清晰的思路，每张规划图都多画几遍，就很容易避免在施工期间增加预算。

A random image with all elevation makes a unique apartment.

四周立面上随机凸出的小房间让公寓显得与众不同。

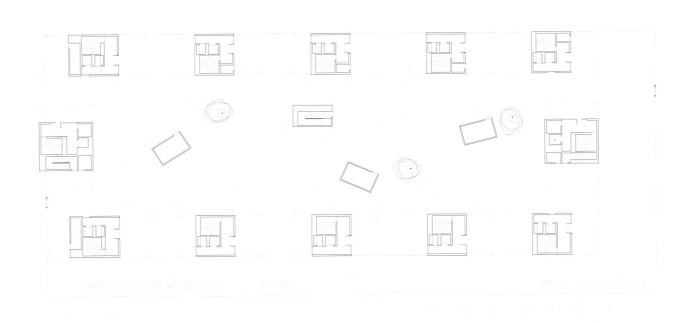

Ground Floor Plan

EXTRA ROOM

BLURRED FRONTIERS

2 FACADES CROSSING SPACE

SERVICES

1. LIVING ROOM
2. KITCHEN
3. MASTER BEDROOM
4. SECOND BEDROOM

0 1 2 3 4 5 m

Typical Dwelling

We were required for small dwellings, with just one or two bedrooms, assigned to young people, what allowed us to think in new housing concepts.

The building plan is a thin strip, with narrow apartments opened onto two different scenes: one side the city, the other the wide private garden. Every unit has a space which crosses from side to side, opening simultaneously to both of them. We seek for dwelling with blurred frontiers between spaces, where inner room shape can be changed, and with the presence of an extra space: a room floating outside the building, attached to the facade, large enough to accommodate any domestic activity.

We approach social housing from present society, where customizing your own house is a way to seek for individuality. In this context, the new room —freely set in the facade— expresses this new understanding: a random image with all elevation being unique, every single dwelling different from each other, and only the intervention of the users helping it to acquire significance.

设计师被要求设计仅带一两个房间的小型住宅，主要面向年轻人，这也促使设计师引入新的住宅设计概念。

公寓的建造方案是建造一套狭长形的公寓，并面向两面不一样的景色：一面是城市，另一面是宽阔的私人花园。每一个单元都包含一个贯通两侧的空间，同时向两侧开放。设计师希望公寓各空间不要划分明显的界限，这样内部房间的形状就可以灵活变。另外，房间还有一个额外空间：一个突出于立面的小房间，附着于立面上，可以提供足够的空间给任何家庭活动。

设计师立足于社会现状所探寻出的方案还能为不同住户的需求打造个性的房间。在这种背景下，自由设置在立面上的房间表达了新的理解：四周立面上随机凸出的小房间让公寓显得与众不同，每间房也都独一无二，只有各自住户方知其特色。

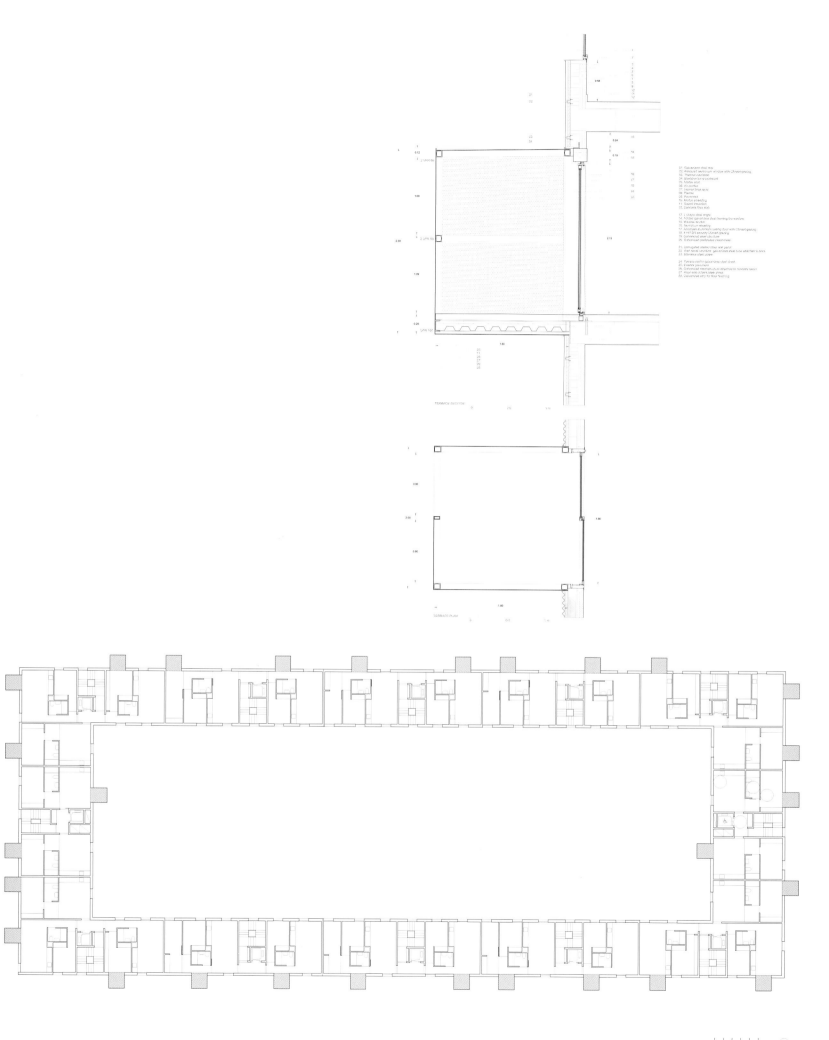

01. Galvanized steel tray
02. Anodized aluminium window with Climalit glazing
03. Thermal insulation
04. Stretcher bond brickwork
05. Mortar seal
06. Air chamber
07. Interior brick layer
08. Plaster
09. Mortar screeding
10. Liquid breakfast
11. Concrete floor slab

12. L-shape steel angle
13. Folded galvanized steel framing the window
14. Window situation
15. Aluminium sheeting
16. Anodized aluminium swing door with Climalit glazing
17. 4+4/12A security Climalit glazing
18. Galvanized steel structure
19. Galvanized perforated metal sheet

21. Corrugated coated steel seal panel
22. Wall panel structure galvanized steel tube attached to brick
23. Stainless steel screw

24. Terrace roof in galvanized steel sheet
25. Granite pavement
26. Galvanized steel structure attached to concrete beam
27. Hoja-sole in bent steel sheet
28. Galvanized strip for floor finishing

TERRACE SECTION

TERRACE PLAN

| | | | | | | N |
| 0 | 1 | 2 | 3 | 4 | 5m | |

Typical Floor Plan

CVETKOVA
APARTMENTS

CVETKOVA 公寓

Architect: Studio Kalamar
Client: Municipality of Murska Sobota
Location: Murska Sobota, Slovenia
Built Area: 6,370 m²
Status: Completed
Units: 57
Photography: Miran Kambič, Studio Kalamar

设计公司：Kalamar 工作室
客户：穆尔斯卡索博塔市政府
地点：斯洛文尼亚穆尔斯卡索博塔市
建筑面积：6 370 m²
状态：已建成
户数：57
摄影：Miran Kambič、Kalamar 工作室

The original construction cost is €5,800,000 that means: construction cost in China is

¥ 13,647,571

per unit cost is ¥ 239,431

原建造成本为 5 800 000 欧元，国内建造成本约

¥ 13 647 571 元

国内平均每户造价约 239 431 元。

Apartment buildings are located in a redeveloped downtown area that has recently been converted from derelict industrial to public and residential use. Despite its central position in the city, the site is removed from main roads and downtown bustle. Next to the new music school on the north of the city block, three apartment buildings rise from a green surface. Their positions reflect the heterogeneous surroundings; variety of directions influences the varying orientations of the volumes.

本案坐落在斯洛文尼亚一个重建的市区，这里曾是废弃的工业区，后被改造为公共和住宅用地。公寓虽然位于市中心，却能免受公路噪声和城市喧嚣的影响。项目共有 3 栋公寓楼，周围绿意盎然，北面是新建的音乐学校。这些公寓的布置反映了周围不同的环境；每套公寓朝向不同方向，展现了不同的立面。

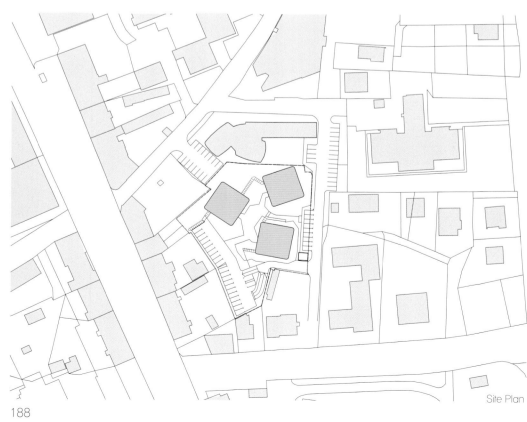

Site Plan

Cost and Creativity 成本与创意

Project's budget was controlled with the implementation of compact design (good ratio between the facade and saleable area) and reasonable use of established construction and material principles.

项目的预算控制与其紧凑的设计（优美的外观和实用面积之间的比率）和合理利用既定的施工方式，以及材料选择有关。

Mosaics in red and grey tiles compose a unique façade in round-angles.

由红色和灰色小瓷砖组成的马赛克立面让这栋圆角建筑颇显独特。

As there are no parallel facades, each apartment can establish an individual character without obstruction. Lower floors comprise four apartments each, top floor is divided into three apartment units, all with flexible floor plans. As a result, long facade surfaces ensure each apartment abundant natural light. The smooth facades with enclosed loggias enfold the volumes into a shimmering ceramic coat, varying in colour tone and reflectivity.

这些建筑中没有两个平行的立面，每一个公寓都单独呈现出自己独特的个性。低层设置四个房间，高层是三个房间，都是不同的户型。长长的立面确保充足的阳光照射。建筑表面是光滑的瓷砖，色调和反射度变化各异。

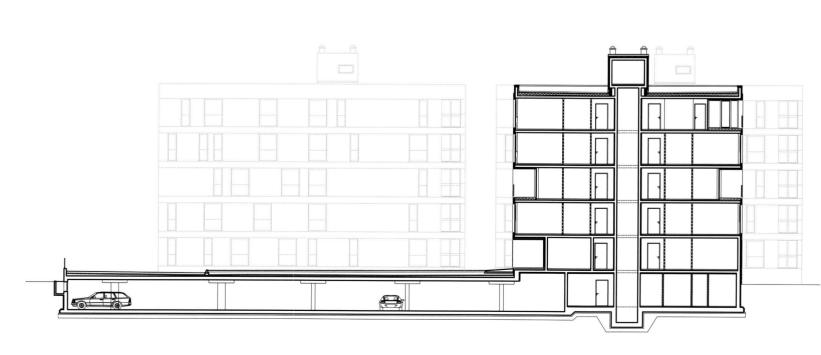

Ground Floor

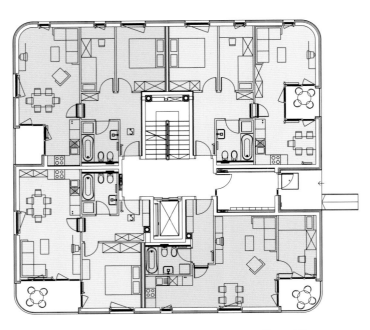

Ground Floor

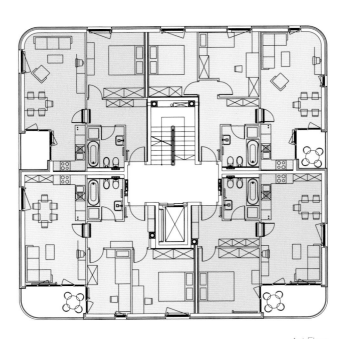

1st Floor

Transparent fence around the complex prevents exclusion from surrounding city life while allowing unsupervised children's play. Gently sloped green areas provide various opportunities for that, and a small playground for the youngest children is located in the eastern corner of the site.

Parking and storage spaces are arranged below ground, serving as a base for all three volumes. Special attention was accorded to energy use of the complex. Each building has a separate boiler room, cooling is arranged individually, the compact facade and energy supply scheme both stimulate economical consumption and thus reduced energy use.

透明的围栏既能让公寓和周围的城市生活相融合，又能保证儿童在无人看管下也能安全游玩。减缓的绿地为各种社区活动提供场地，在东边的角落还有一块供幼童玩耍的小操场。

停车场和储物室都安排在地下层，为 3 栋公寓共用。能源的使用也经过了特别的考虑。每栋公寓都有独立的锅炉房和独立的制冷设备，简洁的立面和能源供应规划能刺激经济消费，从而减少能源的使用。

CRANFIELD UNIVERSITY CHILVER HALL

克兰菲尔德大学 CHILVER 学生公寓

Architect: Stanton Williams Ltd
Client: Cranfield University
Location: Cranfield, U.K.
Site Area: 2, 962 m²
Status: Completed
Units: 106
Photography: Peter Cook

设计公司：Stanton Williams Ltd
客户：克兰菲尔德大学
地点：英国克兰菲尔德市
占地面积：2 962 m²
状态：已建成
户数：106
摄影：彼得·库克

The original construction cost is £4,000,000 that means: construction cost in China is

¥ 9,253,970

per unit cost is ¥ 87,301

原建造成本为 4 000 000 英镑，
国内建造成本约

¥ 9 253 970 元

国内平均每户造价约 87 301 元。

Cranfield University is located on a former air base amid rolling countryside to the south-east of Milton Keynes. Founded after the Second World War as a college of aeronautical engineering, it has grown to specialise in a range of strategic and applied subjects, from technology to management. Having already masterplanned improvements to the campus and designed new postgraduate housing to positive reception, Stanton Williams were recently commissioned by the university to provide accommodation for a further one hundred students. The design has been conceived in response to an exceptionally tight budget and short timescale. Nonetheless, as with our previous work at Cranfield, it aims to rise above the typical standard of new student housing by generating a very real sense of place and community through the considered use of high-quality materials and a contextual response to its site.

克兰菲尔德大学位于前空军基地——米尔顿·凯恩斯的东南方。作为二战后建立的航空工程大学，它已经成长为专注于一系列战略和应用学科、从技术到管理方面的专业学校。在改善校园和设计新的研究生住房提高接纳功能已经取得积极成果时，斯坦顿·威廉姆斯最近被校方要求再提供一百名学生的住宿。设计在短期预算异常紧张的情况下已经开始，就像设计师以前在克兰菲尔德的项目一样，它通过用高质量的材料建成非常实用的社区，使新学生的住房标准不致降低。

Cost and Creativity 成本与创意

The existing buildings on the site prompted a strongly contextual approach, while the use of a cost-effective timber frame to speed construction further presupposed a certain directness.

场地原有的建筑对环境有一定的影响，而用低成本的木构架可以加快建设过程，方便快捷。

The new blocks are therefore clad in red cedar whose organic, soft texture contrasts with the brick of the houses and relates to the domestic nature of the landscape.

新公寓的红色雪松木纹理清晰，质地柔软，与周围房屋的砖瓦和环境形成了鲜明对比。

The site of the new accommodation lies between the university's sports ground and a series of older houses which address an east-west road. Our response diminishes the importance of this road by creating a new north-south axis and threading new paths through the site. We have defined new courtyards at the rear of the existing houses by locating some buildings at right angles to them and others parallel with them. The landscaping of these courtyards has been sensitively handled to generate a sense of "place": they are conceived as spaces where students will wish to linger, to work and to socialise. Their prevailing treatment has been inspired by the residual traces of orchards which once occupied the site, and so "rooms" will be defined within the courtyards by fruit trees and hedgerows. Vehicle access will be possible for delivery and the visual impact of the limited parking areas will be diminished by the use of permeable surfaces surrounded by resilient alpine plants. Benches at building entrances offer a chance to pause at the transition between indoor and outdoor space.

该新公寓位于大学体育场和一片老房子之间，朝向一条东西向的路。设计师创建新的南北轴线，修建一条南北走向的新路通过该区，降低了这条路的重要性。在现有房屋的后部修建新的庭院，确定一些建筑和它相对应或平行，谨慎处理庭院的景观，形成一种"场地"的感觉：学生将它作为逗留、学习和社交的空间。设计灵感来自曾经生长在这里的果园，所以"空间"将被定义在庭院果树和灌木篱墙之内。车辆可驶入有限的停车区域，它被高山植物包围。建筑物入口处长凳给进进出出的行人短暂休息。

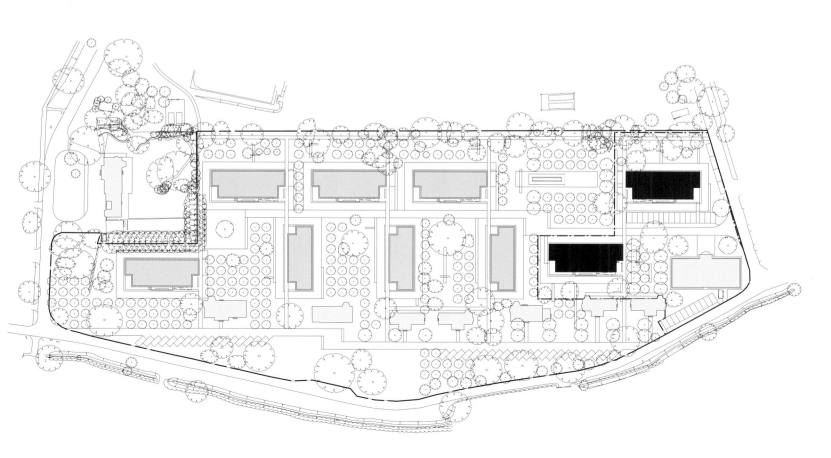

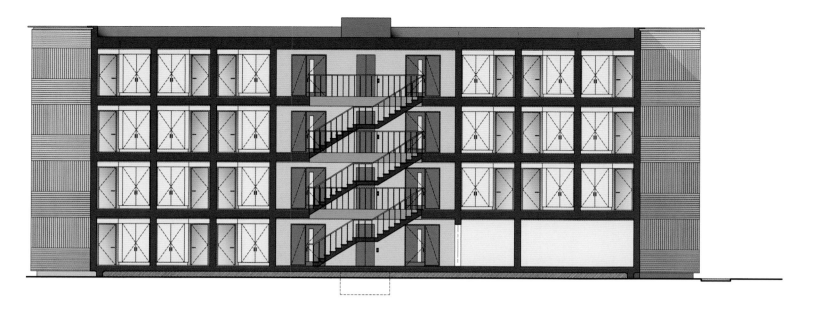

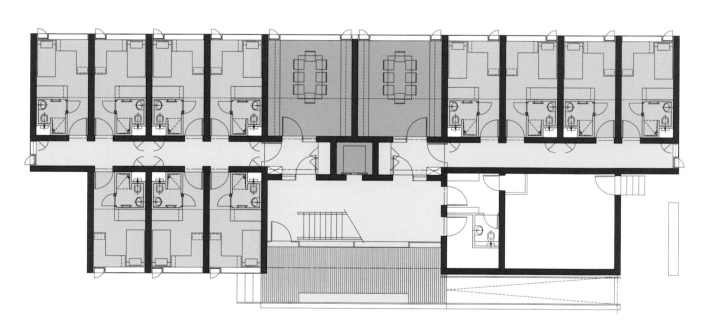

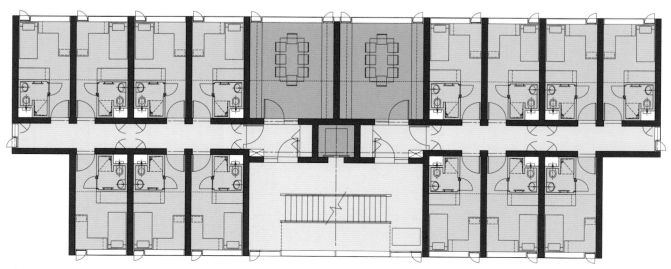

The retained housing along the southern edge of the site prompted a strongly contextual treatment of the new buildings, one which did not attempt to compete with what was there already. Our additions are essentially simple linear forms, though with recessed corners on one side. Those along the northern edge of the site are of four storeys in order to define the boundary with the sports field, while the remainder have three storeys in deference to the existing two-storey houses. The use of a timber frame – essential in allowing us to construct the buildings within the stipulated nine-month period – also presupposed a direct, simple expression.

As with our earlier work at Cranfield, views out from these buildings were essential to their conception. Corridors are terminated by large windows, allowing residents and visitors to define themselves in relation to the surrounding landscape in that the position of the blocks relative to their neighbours is calculated to retain open vistas.

Though good value and efficient construction have been important considerations throughout, the results will, when completed in summer 2009, provide students with an inspiring home that makes the most of their attractive setting.

地块南边界留下的旧房屋为新建筑物提供了很好的借鉴,当然新建筑也不想与旧房屋"竞争",设计师在地铁角落添加的只是些简单的线性形式,沿地块北边界是四层楼建筑与运动场的分界,其余的三层楼建筑则与旧二层相对应。网架的使用直接简单,使公寓在几乎不可能的九个月时间内竣工。

像设计师在克兰菲尔德早期的工作一样,纵观这些建筑,其基本理念都一样。走廊的终端是大窗口,居民和访客可在此欣赏周围景观,与邻居共话远景。

高效的施工也一直是设计师所关注的重点,本案于 2009 年夏竣工,为学生们提供了舒适的公寓,使他们中的大多数有了一个好住处。

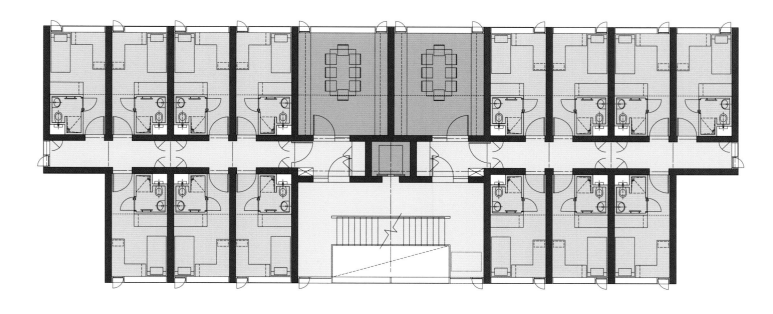

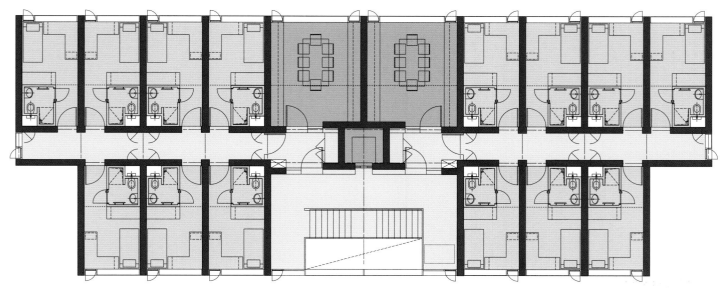

HABITAT
15

HABITAT 15 公寓

Architect: Predock_Frane Architects
Client: Habitat LLC
Location: Hollywood, CA, U.S.A.
Site Area: 3,623.22 m²
Status: Completed
Units: 15
Photography: Predock_Frane Architects,
Elon Schoenholz

设计公司：Predock_Frane 建筑事务所
客户：Habitat LLC
地点：美国加州好莱坞
占地面积：3 623.22 m²
状态：已建成
户数：15
摄影：Predock_Frane 建筑事务所、
Elon Schoenholz

The HABITAT 15 project is a four story, 15-unit infill housing project at the foot of the Hollywood Hills – 1/2 block West of La Brea, and North of Fountain Avenue. Each unit is between 1200 and 2000 square feet. The project is divided into two separate buildings – West and East, with a central courtyard acting as both a buffer and connection between them.

本案分为四部分，共有 15 套单元住房，位于好莱坞山脚下，拉布雷亚以西，喷泉大道以北。本案分为两栋独立的建筑，每个单元在 111 m² 到 186 m² 之间，以一个中央庭院作为东部和西部的缓冲区和它们之间的连接。

Cost effeciency is acheived both by maintaining simple exterior shells with more intricate and varying sections, and by deploying repetitive units of glass and openings.

简洁的立面可以达到控制成本的效果，再以复杂多变的空间做补充，如重复的玻璃窗和开口等。

Windows in different sizes on the wall contribute to an interesting façade with abundant natural lighting.

开满窗户的公寓立面既有趣，又能保证丰富的自然采光。

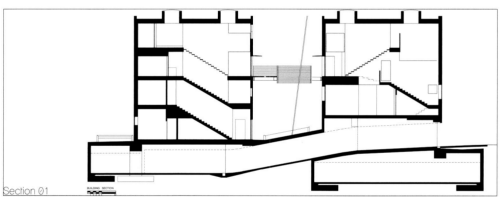

Section 01

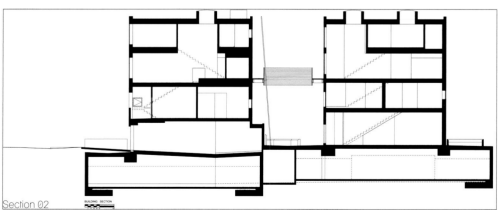

Section 02

Working within the context of a for-profit MUR creates a particular set of parameters; programmatic, physical, and client/cost informed. Our solution is two 4-story buildings placed parallel to the street. Between them is a courtyard, reached via a tunnel through the first building, accommodating shared public space and circulation. The simple cubic shape of the buildings allowed for greater project resources to be deployed to the sectional relations within the units. The separation into two buildings allows half of the unit's access to light on three sides. The townhouse configuration further allows a maximum number of units to be located on the upper levels with greater access to view and light, while the 5 lower level units have exaggerated ceiling heights and open on all sides to the adjacent outdoor spaces. Light penetration is further enhanced in the upper units via room scaled skylights that channel the sky deep into the units. This vertical overlapping of spaces creates complex light play and enlarges the sense of territory occupied by each unit.

本案是赢利性机构的工作场所，能满足各种不同需求，音乐表演、健身、客户咨询等，解决方案是沿街建立四层楼建筑，这些建筑之间是个院子，经由隧道通向一层楼，兼顾空间共享。简洁的立方体外形可更好地利用资源，分成两部分的建筑可使多数楼面三方向采光，连排别墅的造型使大部分单元的上部得到很好的日照。下部五层的净空都很高，全向室外敞开。上部单元中层可让日光照进房间内，让居住者在室内也能感受到外面广阔的天空，竖直的空间增强了光线效果，并增添了建筑的神秘感。

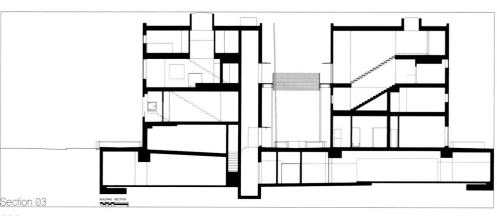

Section 03

Floor 01

Floor 02

Spatial separation is further augmented by orienting the public spaces of the Detroit facing building towards the street while giving the second building 'visual ownership' of the upper courtyard, separated from the first by a bamboo screen. The varying size (based on client standards) and location of the exterior enhances the sense of separation by avoiding overlapping views between units while simultaneously extending views beyond the site. Conversely from the exterior they deny the unitized tendency of MUR's, giving the entire building a singular identity.

The building utilizes double wall construction, vertical heat stack effect cooling via the multi-story volumes, abundant natural lighting, rainwater catchment and redeployment, and environmentally harvested materials.

底特律式外形的临街建筑空间分隔进一步增大，基于客户各种要求，大小和外形的变化增强了距离感，避免了两单元间的视觉重叠，延展了小区内的视野。设计师通过外形的变化，避免了单一化的趋向，使整体建筑更有特征。

建筑采用双层墙结构，垂直热效应得到多重冷却，丰富的自然采光、雨水收集和循环利用，使环境资源得到充分的利用。

ARCHSTONE POTRERO

阿奇斯通·波特雷罗公寓

Architect: David Baker + Partners Architects
Location: San Francisco, California, U.S.A.
Site Area: 174,740 m²
Status: In design
Units: 468

设计公司：大卫·贝克及合伙人建筑事务所
地点：美国加州旧金山
占地面积：174 740 m²
状态：设计中
户数：468

Two buildings with residential units atop flex and retail spaces bring an amenity-rich new neighborhood to the Potrero area. Located on a triangular site at the base of Potrero Hill, the development includes a new 40,000 sf park and is further open to the community by a public mid-block pedestrian mews lined with active uses. Twenty percent of the nearly 500 units will be affordable rental apartments.

本案的两栋公寓楼的下层为商场，上层是住宅，将会是 Potrero 区新型完善的社区建筑。这个坐落于 Potrero 山脚下的三角洲的项目将成为一座占地 3 716.12 m² 的公园，并以半封闭的多功能公共长廊作为连接，向社区开放。公寓共有近 500 间套房，其中的 20% 将作为经济房出租。

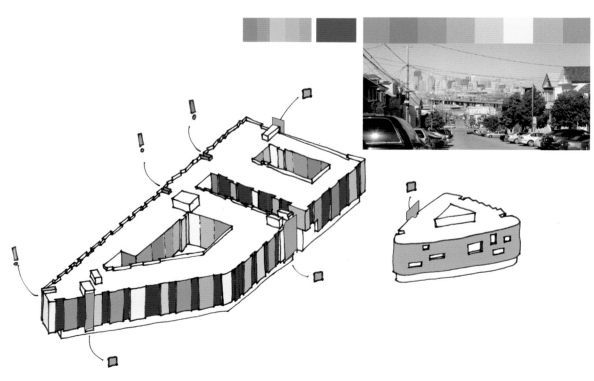

Cost and Creativity 成本与创意

The project brings 15,000 square feet of commercial retail space and approximately 7,000 square feet of PDR/SEW space to provide restaurants and useful services for the larger neighborhood. PDR (production, distribution, and repair) zoning allows for small businesses that fill daily uses. SEW (small enterprise workplaces) zoning allows businesses ranging from 500 to 2,500 sf.

本项目将有 1,393.55 m² 的商业空间和约 650.32 m² 的 PDR（生产、分配、维修）区和 SEW（小型工作室）区，为此大型社区提供餐饮等各种服务设施。PDR 区可供小型企业白天使用，SEW 区则会为不同企业需求提供 46.45 m² 到 232.26 m² 不等的空间。

The flatiron building will occupy the smaller site at the point of the triangle with a vivid and shining facade.

这座熨斗形的公寓楼只占据三角洲的极小部分面积，其色彩鲜明的立面在阳光下熠熠生辉。

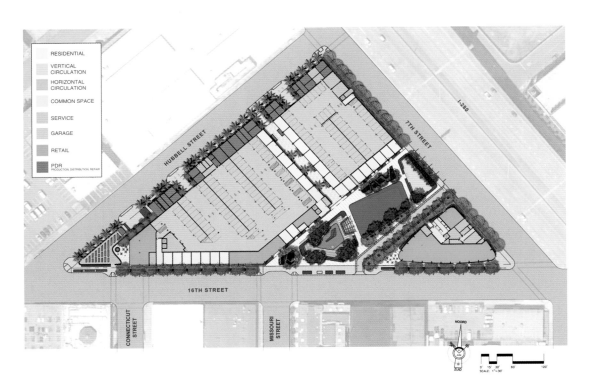

RESIDENTIAL
VERTICAL CIRCULATION
HORIZONTAL CIRCULATION
COMMON SPACE
SERVICE
GARAGE
RETAIL
PDR
PRODUCTION, DISTRIBUTION, REPAIR

HUBBELL STREET

7TH STREET

I-280

16TH STREET

CONNECTICUT STREET

MISSOURI STREET

NOORD

ZUID

0' 15' 30' 60' 120'
SCALE: 1" = 30'

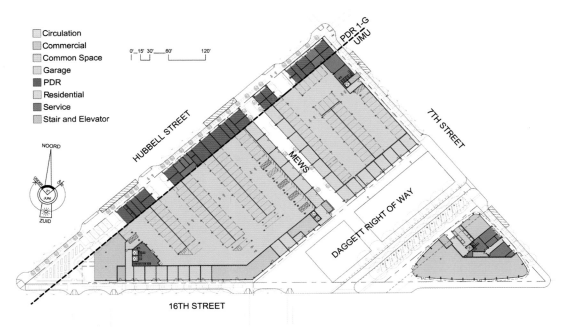

Circulation
Commercial
Common Space
Garage
PDR
Residential
Service
Stair and Elevator

0' 15' 30' 60' 120'

NOORD

ZUID

HUBBELL STREET

MEWS

7TH STREET

PDR 1-G
UMU

DAGGETT RIGHT OF WAY

16TH STREET

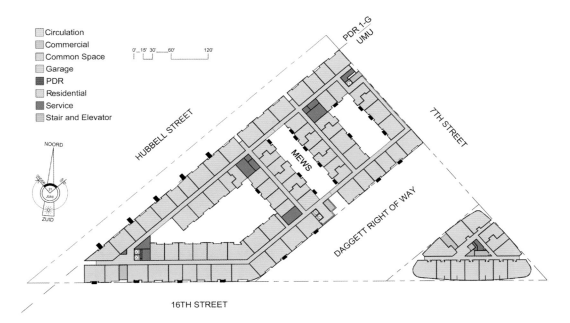

Circulation
Commercial
Common Space
Garage
PDR
Residential
Service
Stair and Elevator

0' 15' 30' 60' 120'

NOORD

ZUID

HUBBELL STREET

MEWS

7TH STREET

PDR 1-G
UMU

DAGGETT RIGHT OF WAY

16TH STREET

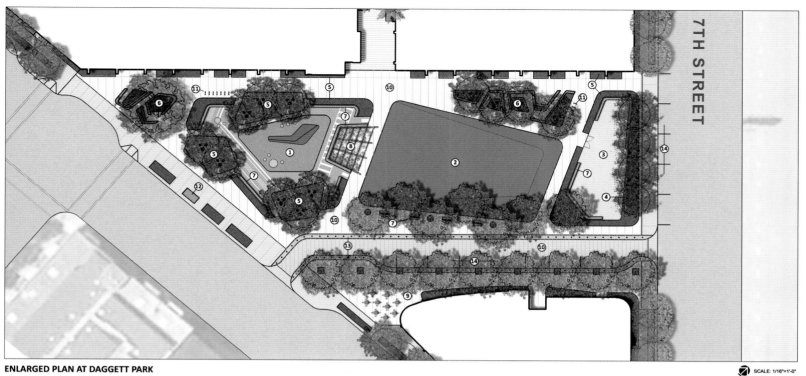

ENLARGED PLAN AT DAGGETT PARK

SCALE: 1/16"=1'-0"

① Play Area - with play area surfacing and play features
② Lawn with Gentle Landform
③ Dog Run with 4' Fence and Gate
④ Screen Wall Feature at 7th Street
⑤ Planting/Stormwater Treatment Area (shrubs and groundcover)
⑥ Built-In Seating and Planters - social seating arrangements
⑦ Built-In Seating at Edge

⑧ Arbor Shade Structure - with power and lighting for community events
⑨ Cafe Seating Area
⑩ Concrete Paving
⑪ Bike Parking
⑫ Bus Shelter
⑬ One Way Travel Lane - with flush curb, detectable warning and bollards
⑭ Parallel Parking (11 total spaces)

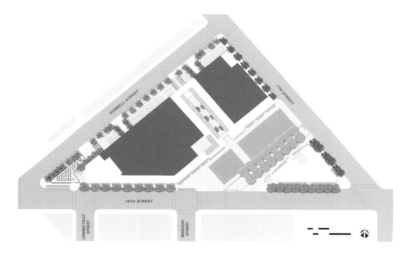

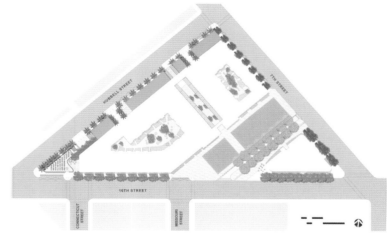

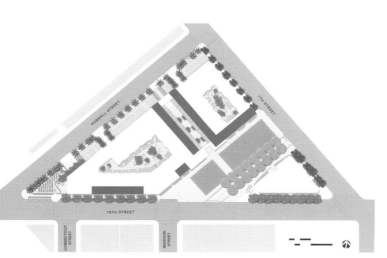

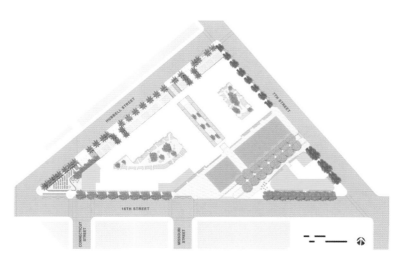

Daggett Park

The former Daggett Street is proposed to be a 1 acre public park designed by CMG Landscape Architects. The park will be owned by the City but permanently maintained by the development. Amenities include a dog walk as well as an event lawn and soft play surface.

There is a smaller public open space planned for the corner of 16th and Hubbell Streets. It is planned to have a community allotment garden as well as outdoor seating for the corner cafe.

Mid-block Mews

At mid-block, a landscaped pedestrian mews is lined with palm trees and active storefronts, as well as a rain garden designed to manage storm water.

Hubbell Park

At the intersection of Hubbell and 16th Streets we plan a park that includes a community allotment garden, outdoor plaza with seating for a cafe, a carshare pod for four vehicles, and a bike station.

At the end of Connecticut Street, Archstone Potrero features a 7,000-sf community garden with raised planting beds. Including this pocket park, the project contains a total of approximately 50,000 square feet of open space dispersed around the project site at grade, within podium courtyards, and in a rooftop garden deck.

达格特公园

达格特街将会被改建成一座占地 4,046.86 m² 的公园，由 CMG 景观建筑事务所负责设计。建成后公园将归市政府所有，由项目开发商进行长期维护。公园里将建成的设施包括一片大型草坪，供游人遛狗或举办活动，还有一处地表柔软的游憩区。

其中有一块小型的公共地被规划出来，作为哈贝尔街和第 16 大道的转角。这里将建造一座社区配套花园和户外咖啡厅。

半封闭长廊

一条半封闭的人行景观长廊，两旁是棕榈树和热闹的店面，并有一座雨水花园以便进行雨水径流的管理。

哈贝尔公园

在哈贝尔街和第 16 大道交会的路口处将计划建造一座社区配套花园、有咖啡厅的露天广场、能容纳 4 辆小车的停车场和自行车存放点。

在康涅狄格街的尽头，Archstone Potrero 公寓还拥有一座 650.32 m² 的社区花园，花园里有高于地面的花圃。包括这座袖珍花园在内，本案总共有 4 645 m² 的空地，分散在场地各处，包括高于地面的庭院和屋顶花园。

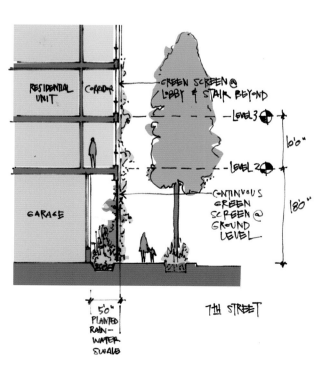

RESIDENTIAL UNIT | CORRIDOR

GREEN SCREEN @ LOBBY & STAIR BEYOND

— LEVEL 3 ⊕ —

16'0"

— LEVEL 2 ⊕ —

CONTINUOUS GREEN SCREEN @ GROUND LEVEL

18'0"

GARAGE

5'0" PLANTED RAIN-WATER SWALE

7TH STREET

40 UNITS HOUSING, FRANCE

法国 40 房住宅

Architect: ECDM
Client: Nexity
Location: Montpellier, France
Built Area: 2,997 m²
Status: Completed 2012
Units: 40

设计公司：ECDM
客户：Nexity
地点：法国蒙彼利埃
建筑面积：2 997 m²
状态：2012 年建成
户数：40

The original construction cost is € 3,800,000 that means: construction cost in China is

¥ 7,284,753

per unit cost is ¥ 182,119

原建造成本为 3 800 000 欧元，国内建造成本约

¥ 7 284 753 元

国内平均每户造价约 182 119 元。

Located between Avenue Pierre Mendes and the Montpellier, the project comes from a will of the municipality to design a garden city near the city center, easily accessible thanks to the numerous infrastructure transportation (major roads, tram, and bicycle network). The Project thus stands in strategic location and exceptional infrastructure, benefits from convenient accessibility to the city center while providing the benefits of an ideal natural environment. The project enjoys surrounding vegetation a unique view to the city and its surroundings.

本案位于法国 Pierre Mendes 大道和蒙彼利埃大道之间，因为市政府计划在市中心建造一座城市花园，所以此处交通设施完善，有主干道、电车、自行车等各种道路。因此，本案所在的位置极为优越，并且设施齐全，交通便利，周边环境良好，绿意葱葱，城市及其周边美景也尽收眼底。

Cost and Creativity 成本与创意

The entire frame includes rational integration of volume, façade, lighting, form, living value and sustainable development.

建筑的整个结构把体量、立面、光照、构筑形式、生活价值和可持续发展等元素合理地搭配在一起，形成一个最优的结合体。

The west and south facades are precast concrete covered with a glaze in pearl white, which sparkles under sunshine.

南、西立面采用了预制混凝土表皮，其表面覆盖了一层珍珠白釉面，在阳光下熠熠生辉。

The starting point of the project is to realize solar achievement. The overall volume is split into two buildings at different heights, and overlapped each other on an L sharp, this provision increasing the surface area of the facades facing south. This volume is based on a precast concrete base common to the three buildings that make up the island and serves as a place of pedestrian access and parking. The volume of the main body is raised, releasing an open space and shelter allowing natural ventilation and lighting. This set of structure strengthens the presence of visible view from the main volume to the surroundings.

本案的出发点是获取阳光。整个体量被分为两部分，各自高度不同并彼此重叠成"L"形，这样可使南立面得到更多的阳光。以预制混凝土做地基，这样可使3栋建筑都在同一地基上，还可作为人行道和停车场。项目的主体量略为提高，这样能留出空间，既可以遮阳，也能促进空气流通和增加自然光照。这样的结构还能扩大建筑的视野范围。

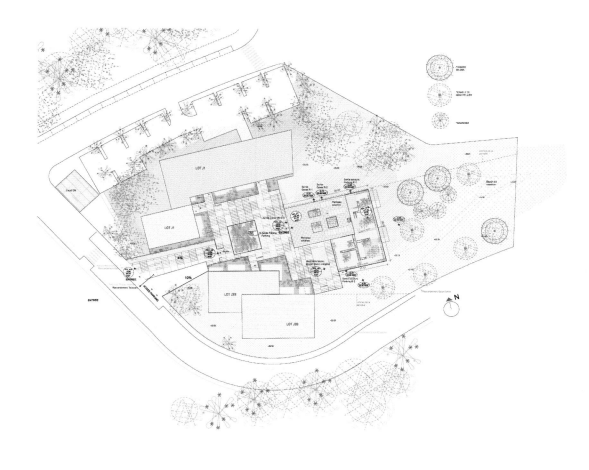

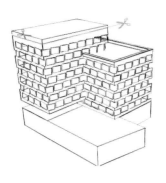
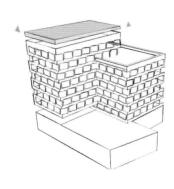
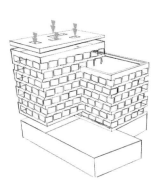

PRINCIPE SUR-TOITURE

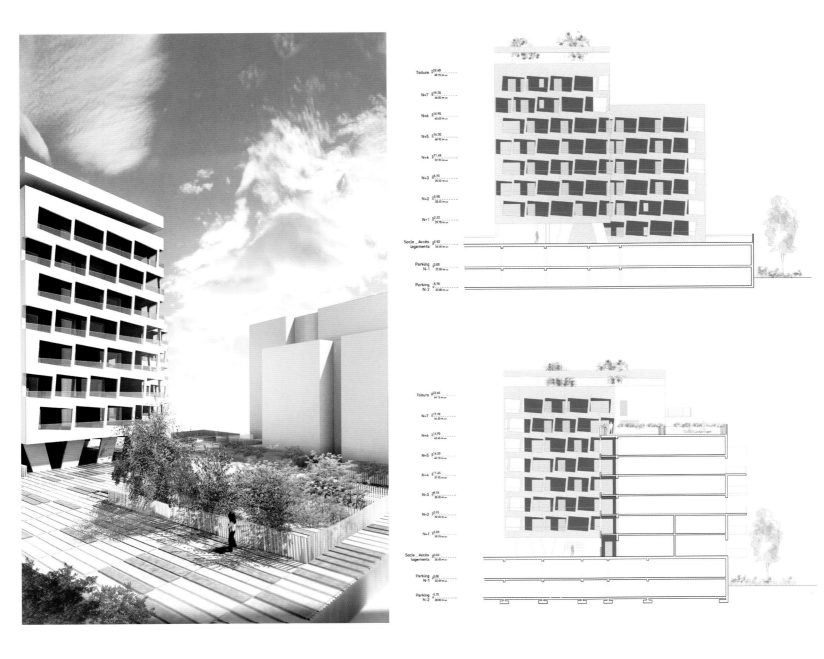

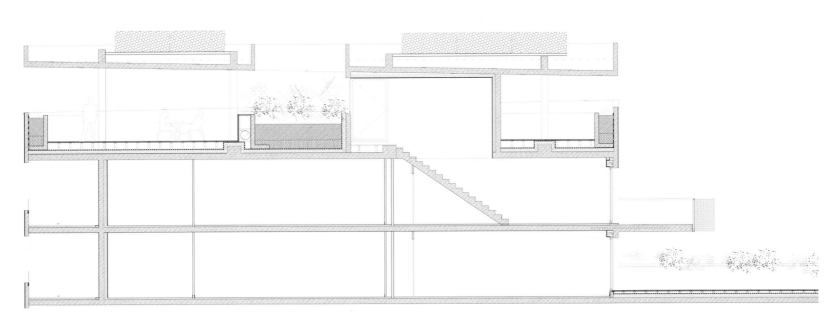

The interior of the loggias is coated dark gray, contrasting with the white facade. This effect intensifies the perception of rhythm and increases the perceived depth of loggias. The north and east facades are covered with a matte white coating, joining the other two facades, accentuates the monolithic while the dissimilarity of the textures and subtly play with contrasts. The materials used for the balconies extend this effect by transparent glass rails for well lighting and ventilation with green concrete panel on the bottom.

内部的凉廊采用暗灰色，与白色的外墙形成鲜明对比，加深了建筑的韵律感和凉廊的深度。北、东立面则是白色的磨砂覆层，和其他两个立面一起突出整栋建筑的纹理，并相互呼应。阳台的材料也延续了这一效果，采用透明的玻璃围栏，光线充足，通风良好，在底部还有一片绿色的混凝土嵌板。

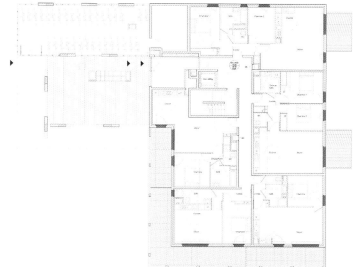

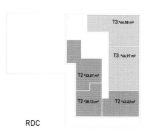

RDC

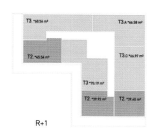

R+1

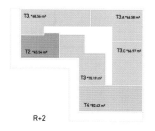

R+2

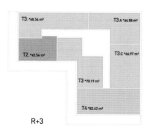

R+3

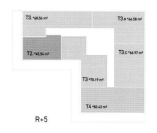

R+4

R+5

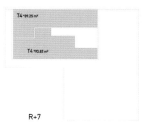

R+6

R+7

COLOR

颜色

SHARING BLOCKS

西班牙共享型公寓

Architect: Guallart Architects
Client: VISOREN RENTA
Location: Gandía, Valencia, Spain
Site Area: 10,174.74 m²
Status: Completed
Units: 143
Photography: Adrià Goula

设计公司：Guallart 建筑事务所
客户：VISOREN RENTA
地点：西班牙巴伦西亚市甘迪亚镇
占地面积：10 174.74 m²
状态：已建成
户数：143
摄影：Adrià Goula

The original construction cost is € 8,000,000
that means: construction cost in China is

¥ 17,604,773

per unit cost is ¥ 123,110

原建造成本为 8 000 000 欧元，
国内建造成本约

¥17 604 773 元

国内平均每户造价约 123 110 元。

This project was developed in Gandía, a town with a population of 75,000 to the south of Valencia. The aim was to develop a hybrid project that would function essentially as a student residence while meeting the requirements of social housing, with the corresponding standards and characteristics. The proposed program includes 102 apartments for young people, 40 apartments for senior citizens, and a civic and social centre for the town council.

本案位于巴伦西亚南部小镇甘迪亚，小镇拥有 7.5 万人口。本案作为学生公寓以及社会住房两大功能的混合体，并具有相应的标准和特点，其中 102 个房间为年轻学生使用，40 个房间是社会住房。

Cost and Creativity 成本与创意

The budget of this project is already fixed before the construction begins. And we control the budget monthly, as they do work, they are payed for that.

施工开始前，本项目的预算就已经被确定下来。设计师对每个月的建设预算都有一定的控制，对工人实行按劳分配薪酬。

Red stripes capture views from a long distance.

鲜红的条纹让人在远处就被吸引。

The most interesting question from a programmatic point of view is the provision of shared spaces in the apartments for young people, which is in effect a new version from the traditional residence for young people. In Spain the national Housing Plan clearly establishes that apartments can be built with an area of between 30 and 45 m², with up to 20% of shared space, but does not specify where or how this should be located.

The fact is that the idea of sharing spaces is fully compatible with the goals of social and environmental sustainability, grounded as it is on the principle of "doing more with less": that is, offering people more resources through the mechanism of sharing. Recent analyses have identified a minimum of thirteen basic functions related to the fact of dwelling. Some of these are clearly private (sleeping, bathing, etc.), while others can have a semi-public or shared nature: eating, relaxing, digital working, washing clothes, etc. These resources can be shared within a single dwelling, between two dwellings, between individuals on the same floor or two adjoining floors, on the scale of a whole building or between different buildings in the same neighborhood.

本案最有趣的部分在于专为年轻人设计的共享空间，这在传统住房中并不多见。根据西班牙法律，30 m² ~ 45 m² 的公寓房间应有 20% 共享部分，但是这 20% 的共享部分在哪里，未曾指出。

其实共享理念与可持续发展是完全相兼容的，以"用较少的能源做更多的事"为原则，人们可以通过共享更好地利用环境资源。设计师分析了 13 个居住基本功能，私人的功能是睡觉、洗澡等，而可以共享和半共享的部分是吃饭、休息、洗衣和电脑办公。这种共享可以存在于一个住宅中，也可以存在于两个住宅中，也可以是同一层楼，或者两个相邻的楼层，甚至一个建筑内或者一个街区的不同建筑物之中。

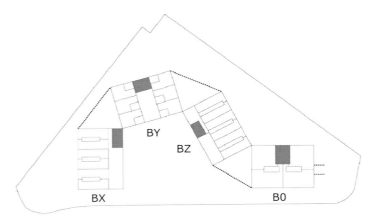

BY

BZ

BX   B0

BLOQUE X
Privado=1728m2
Compartido=288m2

BLOQUE Y
Privado=1296m2
Compartido A=216m2
Compartido B=306m2

BLOQUE Z
Privado=648m2
Compartido=108m2

BLOQUE 0
Privado=648m2
Compartido=108m2

GEOGRAFÍA:
100 jóvenes habitarán un edificio de viviendas universitarias

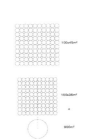

GEOMETRÍA:
Las residencias tradicionale tienen viviendas individuales y servicio

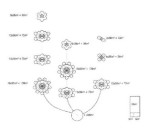

LÓGICA:
Agrupación de viviendas en comunidades con tres grados de privacidad

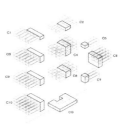

ESTRUCTURA:
Estructura de las viviendas

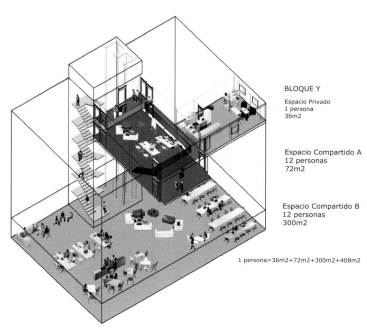

BLOQUE Y

Espacio Privado
1 persona
36m2

Espacio Compartido A
12 personas
72m2

Espacio Compartido B
12 personas
300m2

1 persona=36m2+72m2+300m2+408m2

Espacio Compartido

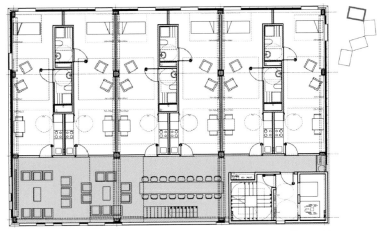

Bloque X

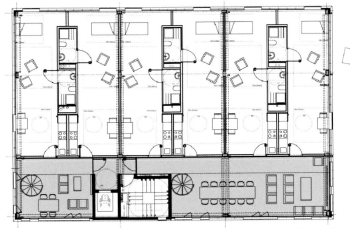

Bloque Z

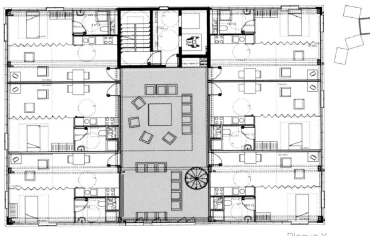

Bloque Y

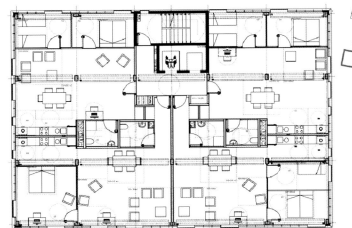

Bloque O

The key, then, is to choose the scale at which we want to share resources so as to create a particular model of habitability or another. If we construct 102 apartments of 45 m² each, which may share 20% of their floor area, we can have up to 918 m² of shared space. This could be in the form of 51 shared spaces of 18 m² (each apartment in a pair contributing 9 m²), or a single space of 918 m². Our proposal puts forward an interesting and innovative model with which to define three scales of habitability: A first, individual scale of 36 m², comprising the kitchen, bathroom and rest area in a loft-style apartment. A second, intermediate scale of 108, 72, 36, 24 and 12 m², shared by 18, 12, 6, 4 or 2 people, on every second floor. This comprises a spacious living area and contact and work areas. A third and larger scale of 306 m², shared by all 102 people and located on the ground floor, which will include a lounge, a laundry, Internet access and a library.

关键则在于实现共享的模式，设计师希望打造多种特定的生活模式，并实现这种共享。102 套 45 m² 的公寓，20% 的共享面积是 918 m²。这 918 m² 可以分为 51 个占地 18 m² 的小共享空间（每间公寓贡献出 9 m²），或者直接将这 918 m² 作为一个整体使用。设计师提出了一个有趣又创新的模式，可提供 3 种不同的生活方式：个别规模的 36 m² 的公寓，有厨房、卫生间、休息区。第 2 部分的公寓房分别有 108 m²、72 m²、36 m²、24 m² 和 12 m² 的 18 人、12 人、6 人、4 人、2 人共享的区域。这里有宽敞的客厅和工作区域。第三部分是一个位于底层的 306 m² 休息室，能供 102 人共享，并配有洗衣房，上网区和图书馆。

RICHARDSON APARTMENTS

理查森公寓

Architect: David Baker + Partners Architects
Associate Architect: Baker Vilar Architects
Client: Community Housing Partnership, Mercy Housing California
Location: San Francisco, California, U.S.A.
Site Area: 18,906 m²
Status: Completed
Units: 120
Photography: Bruce Damonte

设计公司：大卫·贝克及合伙人建筑事务所
合作公司：贝克·维拉尔建筑事务所
客户：Community Housing Partnership, Mercy Housing California
地点：美国加州旧金山
占地面积：18 906 m²
状态：已建成
户数：120
摄影：布鲁斯·戴蒙特

The original construction cost is $ 26,860,000 that means: construction cost in China is

¥ 29,540,417

per unit cost is ¥ 246,170

原建造成本为 26 860 000 美元，国内建造成本约

¥ 29 540 417 元

国内平均每户造价约 246 170 元。

Formerly a parking lot on the southeast corner of Fulton and Gough streets, the Drs. Julian + Raye Richardson Apartments has risen on one of the sites freed for development by the demolition of the collapsed Central Freeway. This five-story building will provide permanent supportive housing for a very-low-income, formerly homeless population. The project is part of the Market + Octavia Neighborhood Plan, which aims to create a dense transit-oriented neighborhood with housing over retail and streets that are friendly to pedestrians and bicyclists.

理查森公寓位于富尔顿和高夫路的东南角，这里原本是一座停车场，在崩塌的中央高速公路被拆除后，空出的其中一片土地就被用于公寓的建设。建成的公寓高 5 层，将为低收入和无家可归的人群提供永久性的保障性住房。本案是 Market + Octavia 社区计划的一部分，其目的是要建造一个以密集交通为导向的社区，底层是零售店和道路，上层是住房，行人和自行车可随意通行。

RICHARDSON APARTMENTS

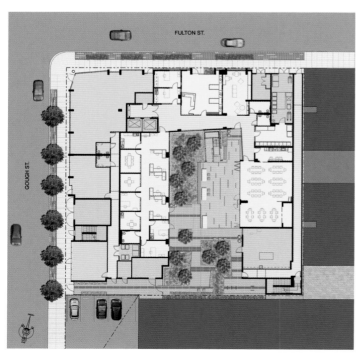

The building responds to the existing fabric of the neighborhood by varying the colors, materials and heights on its façade to suggest a collection of more modest related structures. A prominent corner bay rising over the retail entryway at Fulton and Gough contributes to the dramatic view down the length of the building that culminates with City Hall in the near distance.

为了与社区周围的建筑保持和谐一致，公寓的立面通过不同的外观色彩、材料和高度来表达其毫不张扬的构造。商场的入口位于富尔顿和高夫路的转角处，从高起的地形向下看去，公寓的突出造型颇为醒目，并和附近的市政厅交相辉映。

Cost and Creativity 成本与创意

The project implements various energy saving measures to significantly reduce energy consumption.

公寓实行各种节能措施，大幅度减少能量消耗。

The building varies the colors, materials and heights on its façade to suggest a collection of more modest related structures in a striking view.

公寓的立面色彩艳丽、材料多样并且高低错落，以一种极其醒目的视觉冲击效果反衬出建筑低调的结构。

The residents' entrance on Fulton Street features a spacious lobby with a reception station. An open grand staircase connecting the first through fifth floor levels reduces reliance on the elevator and encourages interaction between residents. Beyond the lobby, the south-facing courtyard frames an expansive existing mural – a paintand-glass mosaic of dancers on the side of the Performing Arts garage. Four levels of fully equipped studio apartments sit atop common spaces surround the private landscaped courtyard.

A tall retail level with an awning trellis that extends over the sidewalk helps maintain a human scale at the street edge. One retail space is dedicated to a work-training program for residents. Other supportive services and features include a counseling center and a residents' lounge, as well as a prominent community room. Additionally there is dedicated on-site medical suite reserved for resident care.

The project is being designed and built with the guidance of the Build It Green GreenPoint Rated and Green Communities checklists, with sustainable features such as a purifying bioswale in the court, sunshades, and possibly solar electric and domestic-hot-water panels. In keeping with the intention of a dense, transitoriented neighborhood, there is no on-site car parking, and bike parking facilities are provided.

公寓大门面向富尔顿路，走进公寓便是设有接待处的宽敞的大堂。大气宽阔的大型楼梯连通一楼到五楼，以减少住户对电梯的依赖，鼓励人与人之间的交流。大堂外是一座朝南的庭院，一幅由油漆和玻璃马赛克组成的大型舞者壁画颇为引人注目。公寓里还有 4 层设备齐全的工作室，位于被私家园林包围的公共空间的上层。

大型商场遮阳棚延伸出人行道，可为路边行人遮阳挡雨，保持人流量。商场里还专门为居民设置了工作培训部门和其他配套服务及功能设施，如辅导中心、休闲区和交流室等。此外还有专业的现场医疗部门，为居民提供及时的医疗服务。

本案在设计和建造过程中都以"建造绿色"组织的"顶级绿色奖"和绿色社区的标准作为指导原则，强调可持续发展功能，如庭院里的生态净化、遮阳棚、太阳能电热水器等。考虑到这是一座以密集交通为导向的社区，公寓没有设置停车场，只提供自行车的停放设施。

8TH + HOWARD FAMILY APARTMENT + SOMA STUDIOS

霍华德街 8 号家庭式公寓及 SOMA 办公楼

Architect: David Baker + Partners Architects
Associate Architect: I.A. Gonzales Architects
Client: Citizens Housing Corporation
Location: San Francisco, California, U.S.A.
Site Area: 16,432.13 m²
Status: Completed
Units: 162
Photography: Bill Owens, Brian Rose

设计公司：大卫·贝克及合伙人建筑事务所
合作公司：I.A. 冈萨雷斯建筑事务所
客户：Citizens Housing Corporation
地点：美国加州旧金山
占地面积：16 432.13 m²
状态：已建成
户数：162
摄影：比尔·欧文斯、布赖恩·罗斯

The original construction cost is $ 23,877,990, that means: construction cost in China is

¥ 26,260,825

per unit cost is ¥ 162,104

原建造成本为 23 877 990 美元，国内建造成本约

¥ 26 260 825 元

国内平均每户造价约 162 104 元。

The city corner comes alive with 8th + Howard/SOMA Studios' undulating edge, bright geometric mural, bustling corner market, and hand-crafted glass and steel gate. An eclectic mix of artists, immigrants, veterans, and young people have made these 162 units of affordable housing their home in San Francisco.

波浪形的立面、明亮的几何壁画、热闹的转角市场，以及手工制作的玻璃钢闸门，本案的建成为城市的一角带来了新的生机，为旧金山不同的人群，如移民、退伍军人和年轻人等提供了 162 套经济住房。

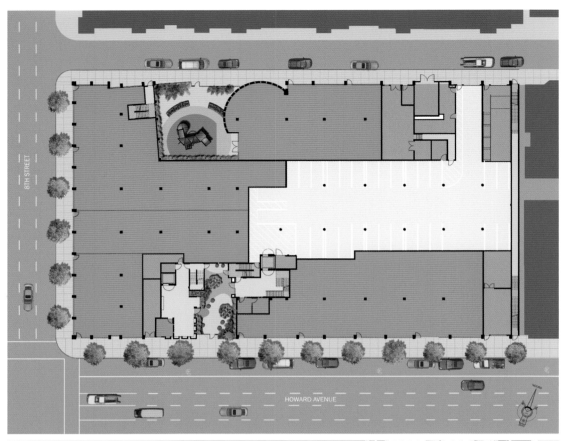

Unbundled parking frees up space for a child-care center and 19,000 square feet of neighborhood-serving retail, including an organic market. The building is visually rich, with a modern patchwork mural punctuated by long windows on one side, a sinuous wall creating curving rooms on the other. The careful, yet playful, design allows for a depth of detail uncommon in affordable housing.

分散的停车场为儿童保健中心和占地 1 765.16 m² 的社区零售店提供更多的空间，其中还包括一个有机市场。公寓一侧的彩色拼接墙面上以长窗作为点缀，另一侧的弧形墙面则创造出半圆形的房间，给人带来丰富的视觉效果。如此精致而又不失生动的细节和深入的设计在经济房项目中并不多见。

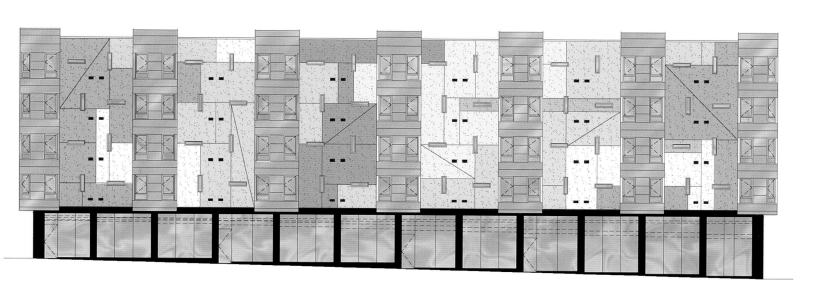

Cost and Creativity 成本与创意

A combination of office and apartment meets the different requirements of users, and make the allocation of resources more rational and optimized.

办公楼和公寓楼的组合满足了更多不同用户的要求，使资源配置更为优化。

Bright and dark colors compose the front and back façade contrastingly while harmoniously.

亮色和暗色构成了建筑的前后立面，对比鲜明却又不失和谐。

This five-story complex is divided in two: The studio side houses modern single-occupancy studios with private baths; the family side features affordable family apartments. Both sides enjoy private courtyards and a wealth of community services.

The gap in the urban edge provides a view into the softer interior open space, sharing it with the civic whole. Entrances to the housing are through semipublic landscaped courtyards that also serve as outdoor green space for public functions.

这栋 5 层高的大楼分为办公楼和公寓两部分：办公楼部分是现代风格的单房，配有私人浴室；公寓部分则是为家庭服务。这两部分都享有私家庭院和完善的社区服务。

在街道外部顺着楼宇之间的空隙往里看，内部柔和的开阔空间随即映入眼帘，似乎在邀请市民们前来分享。公寓的入口处有半公共的景观庭院，同时也作为公共户外绿地。

MD
HOUSING

MD 公寓

Architect: VAStudio
Associate: Atelier Loza e Gradim
Client: Gaia Social
Location: Vila Nova de Gaia, Portugal
Site Area: 2,660 m²
Status: Completed
Units: 36
Photography: Alberto Plácido

设计公司：VAStudio
合作人：Atelier Loza e Gradim
客户：Gaia Social
地点：葡萄牙 Vila Nova de Gaia
占地面积：2 660 m²
状态：已建成
户数：36
摄影：Alberto Plácido

The original construction cost is € 420 /m²
that means: construction cost in China is

¥ 7,270,091
per unit cost is ¥ 201,947

原建造成本为 420 欧元 /m²，
国内建造成本约

¥ 7 270 091 元
国内平均每户造价约 201 947 元。

Located in the historic center, it has a privileged perspective on the Douro River, the caves and the port. Sits on an extreme pitched lot, part of a very irregular urban framework, typical to the site. Regarding these aspects, along with sun exposure, the functional program density and the inherent low cost, this whole thing has resulted in the concentration of all housing typologies , "sliced" by the distinct character of the volume to the west, set for low-cost housing for sale, and the east volume to the PER housing.

本案坐落于历史悠久的市中心，地理位置优越，坐拥杜罗河、洞穴和港口的无限风光。因为场地是极度倾斜的，属于城市框架中极不规则的一部分，所以设计颇具挑战性。考虑到这些要素和日照的方向，公寓具有多功能、低成本的特征，设计师内置了各种不同的户型，以明显的特色划分出东西两部分，西部作为经济房，东部则作为单房。

Cost and Creativity 成本与创意

Effective management of the construction budget was based on a detailed study of the modular structure calculation, and an optimized construction process associated with the use of inexpensive materials, to meet all requirements regarding management and maintenance of the building and its surroundings.

有效的管理施工预算是基于模块化结构的计算，在优化施工过程中，设计师选用物美价廉的材料，使建筑和周围环境的管理和维护要求都能得到满足。

This entire polygon undergoes a twist and shock the context with its striking colors.

设计师对多边形结构进行扭曲，并以此为公寓的结构，用鲜明的色彩照亮了周边的环境。

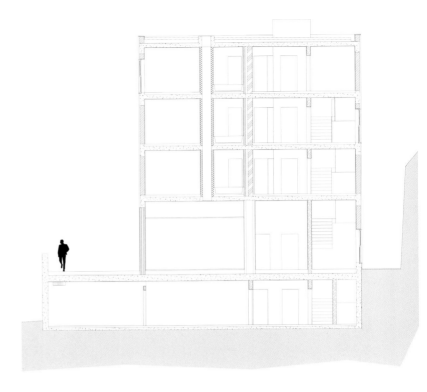

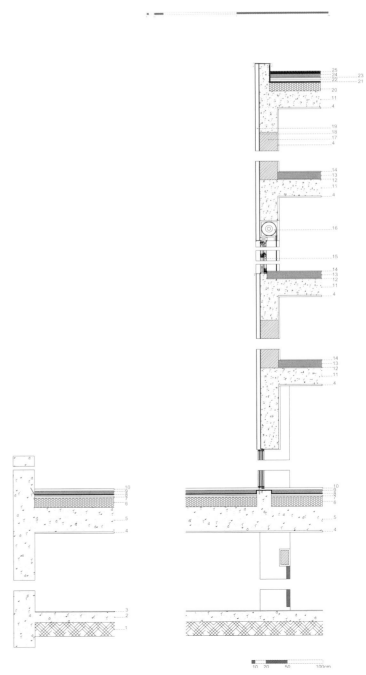

The primary intention of reducing the impact of the volume height was demonstrated by the maintenance of the existing slopes and retaining walls, which are incorporated in the design to relate the number of floors of each body with the height of the existing roads to provide the rear communication among themselves through outdoor galleries.

This entire polygon undergoes a twist, and is based on a leveled public platform that will establish a variety of relationships with the surrounding public space, so it becomes embedded in the urban framework. This platform appears as a groundwork, composed of semi-underground garages, which defines the block and supports this contemporary element in the surrounding historical context.

减轻建筑体量高度的影响主要通过保留原来的斜坡及挡土墙来实现。设计师也考虑到了每栋楼层数和原有街道的高度，并提供户外走廊来加强住户间的交流。

公寓以多边形的结构进行扭曲，建立在水平的公共平台上，可与周围的公共空间建立多种联系，成为城市框架中不可或缺的一部分。以平台为基础，包括半地下车库，公寓将成为周边历史环境中别具一格的现代元素。

DAY CARE AND ELDERLY RESIDENTIAL CENTRE

日间护理中心及老人公寓

Architect: CVDB Arquitectos
Client: Oeiras Municipality
Location: Oeiras, Portugal
Built Area: 5,000 m²
Status: Completed
Units: 60
Photography: FG+SG and Diogo Burnay

设计公司：CVDB Arquitectos
客户：奥埃拉什市政府
地点：葡萄牙奥埃拉什市
建筑面积：5 000 m²
状态：已建成
户数：60
摄影：FG+SG、Diogo Burnay

The original construction cost is € 4,000,000
that means: construction cost in China is

¥ 11,539,826

per unit cost is ¥ 192,330

原建造成本为 4 000 000 欧元，
国内建造成本约

¥ 11 539 826 元

国内平均每户造价约 192 330 元。

The project is located between in an urban area with a residential neighbourhood mainly with social housing for low income families uphill and an area with various public buildings at the lower part of the valley. The elderly day centre is in between these rather different social, programmatic and physical realities, thus celebrating both its private and public character.

本案位于市区，在靠近山谷的高处有一片为低收入家庭准备的经济房住宅区，山谷的低处则是各类公共建筑区，本案就处在这两片区域之间。日间老人护理中心和这些社会性建筑的目的和用途都大不相同，兼具公共和私人的性质。

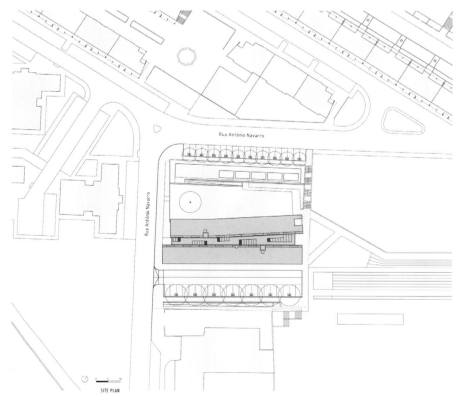

Rua António Navarro

SITE PLAN

The 60 dwellings are set out in two rows facing east-west on four interconnected floors. The spatial sequences of the circulation and of the voids in the central area are part of a bioclimatic strategy which creates several meeting points for people to talk and allow vertical transparency between floors. Conceptually this space will be sufficiently flexible to allow residents to come to feel at home in the place they live in. The ground floor is constructed by black slate volumes as a counterpoint to the transparent large glazed areas which accommodate the building's social and most public spaces. On the outside, the private garden, which is a social meeting point of the inhabitants of the Centre, appears as if were a visual extension of the public space as it can be opened to the public as well. The building/garden seeks to fit in between the various urban tensions that are present in the landscape.

公寓共有 60 个房间，高 4 层，分为两排，分别朝向东、西方，各楼层都相互连接。中间区域的空间序列考虑到空气流通和留空位置，采用了生物气候战略的一部分，提供多个交流区，方便住户的沟通，并给楼层间带来垂直透明感。从概念上讲，即是空间具有充分的灵活性，让居民身处何处都感觉像在自己家里一样自在。首层的黑色岩石体量作为以玻璃围合的大型透明区域的焦点，为公寓提供聚会和活动场地。外部的私人花园也是该公寓的社交区，是公共空间的视觉延伸，同时也可向公众开放。公寓和花园都希望以景观来缓解城市中的各种紧张情绪。

Cost and Creativity 成本与创意

The precise control of building cost is obtained with a continuous analysis of the options that the project can take during its development, following a balanced path that can bring together and have a positive response to the functional, economical and aesthetical issues.

建设成本的精确控制是通过持续分析项目建造过程而得的。功能、经济和美学上的平衡可以为建筑带来积极的影响。

The building presents itself as a counterpoint to the architectural character of the existing housing neighbourhood and therefore it tries to establish strong sense of architectural identity for the existing communities, through its particular coloured and textured materials.

公寓本身作为与街区内现有住房的建筑风格相对应的建筑，试图通过特定的色彩和有纹理的材料，建立其与众不同的建筑感，以和其他建筑有明显区分。

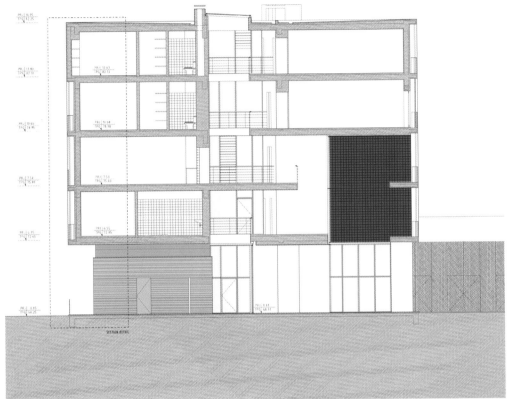

A-A' Section

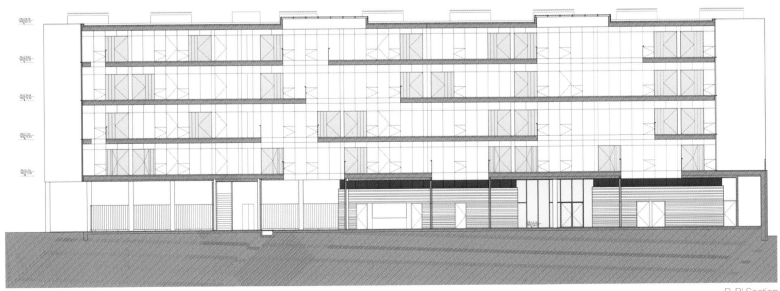

B-B' Section

BIOCLIMATIC BUILDING TREND

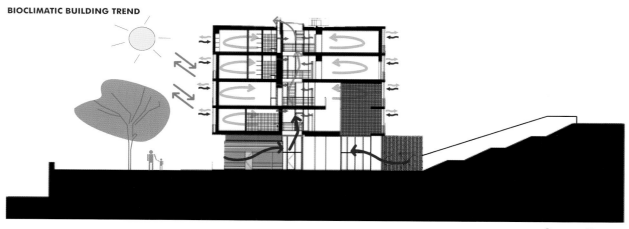

Summer Diagram

VENTILATION BUILDING TREND

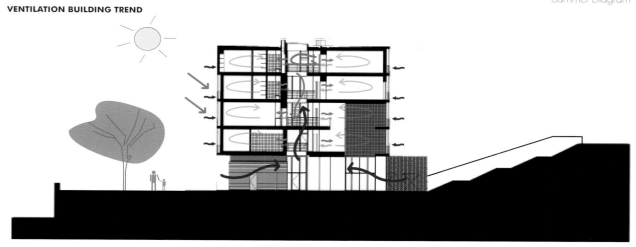

Winter Diagram

SECTIONS APARTMENT A

SECTIONS APARTMENT TYPE B

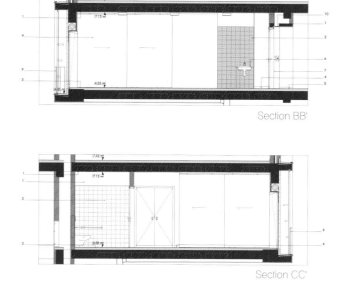

Section BB'

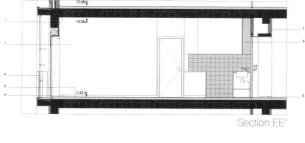

Section EE'

Section CC'

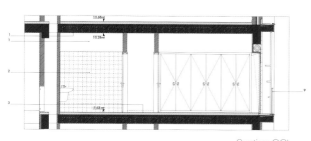

Section GG'

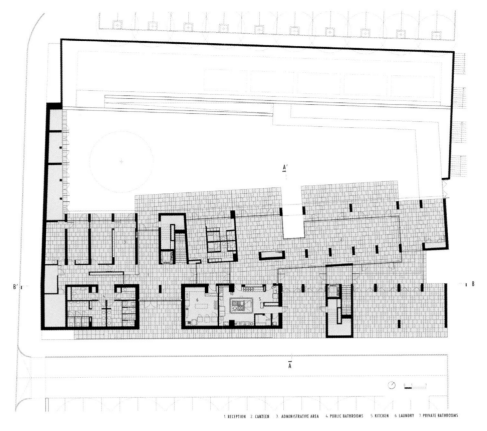

1. RECEPTION 2. CANTEEN 3. ADMINISTRATIVE AREA 4. PUBLIC BATHROOMS 5. KITCHEN 6. LAUNDRY 7. PRIVATE BATHROOMS

Plan Level 00

A'

B'

B

A

Plan Level 02

RUNDESKOGEN

RUNDESKOGEN 住宅

Architect: Helen & Hard
Associate: DRMM
Client: Kruse Bolig AS
Location: Sandnes, Norway
Site Area: 8,800 m²
Status: Completed 2012
Units: 114
Photography: Helen & Hard

设计公司：Helen & Hard
合作公司：DRMM
客户：Kruse Bolig AS
地点：挪威桑内斯
占地面积：8 800 m²
状态：2012 年建成
户数：114
摄影：Helen & Hard

The original construction cost is € 45,139,534, that means: construction cost in China is

¥ 56,894,394

per unit cost is ¥ 499,074

原建造成本为 45 139 534 欧元，国内建造成本约

¥ 56 894 394 元

国内平均每户造价约 499 074 元。

The project is situated in an infrastructural node between three city centers on the west coast of Norway. The region is dominated by single-family houses and small-scale housing projects, creating a context which accentuates the exceptional height and volume of our project. The density and concentration of the project was maximized in order to spare and protect a recently-discovered Viking grave in the neighboring hillside. We were thus able to retain the same number of dwellings, while retaining a respectful distance from the historical relic.

本案坐落于挪威西海岸的三个城市中心之间的基础设施节点上。该地区主要为独立家庭住宅和小规模的住宅项目，这样的环境更突显了本案建筑的高度和体积。本案由该地区的低密度扩大，达到最大的密度，原因是最近在邻近山坡上发现了维京（北欧海盗）墓，这样一来，就可以保留一定数量的原有住房而不影响历史遗迹。

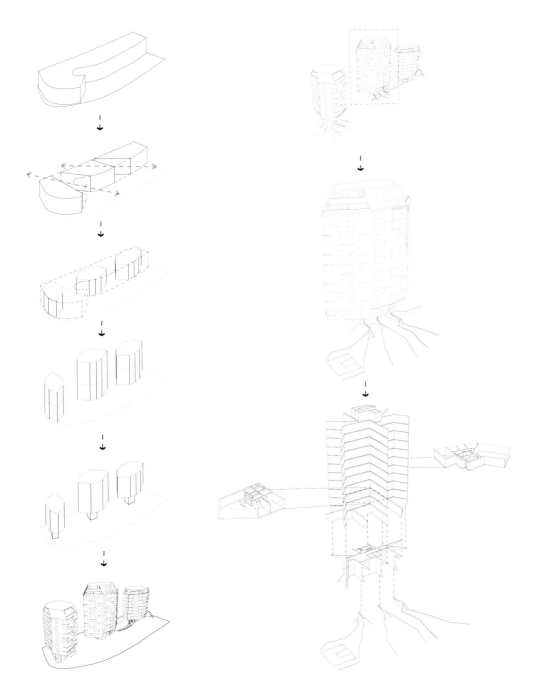

Rundeskogen consists of three towers of 12-15 stories with a total of 114 flats between 60 m² and 140 m². The project is located at a major traffic intersection linking three bigger cities of the region. The emphasis has been to balance the tall building typology with generous and attractive public green space on the ground. To minimize the footprint of the three towers and retain the fjord view for neighbors, the first apartment floors have been lifted off the ground, cantilevering from the core and creating covered outdoor spaces.

本案由三栋塔楼组成，分别高 12 ~ 15 层不等，共 114 个单位，面积从 60 m² 到 140 m² 不等。项目的位置在三个城市中心之间的基础设施节点上。项目的重点是如何在高大的建筑类型和开阔迷人的公共绿地间保持平衡。为了将三栋塔楼的占地面积减到最小，同时确保居住者能欣赏到附近海湾的无限风光，第一栋公寓楼被抬离地面，悬挑的结构提供了带顶棚的户外活动空间。

Cost and Creativity 成本与创意

The "Trunk" structure can effectively resuce occupation of land resources, thereby cutting down costs.

"树干"结构能有效地减少建筑对土地资源的占用，从而降低了成本。

The facades are clad with triangular, metallic plates that reflect the light differently as one moves past, thereby creating an additional effect of movement.

外墙立面覆盖着三角形的复合再生木板材，当人们走过时可反射不同的光，从而创造出一种额外的动态效果。

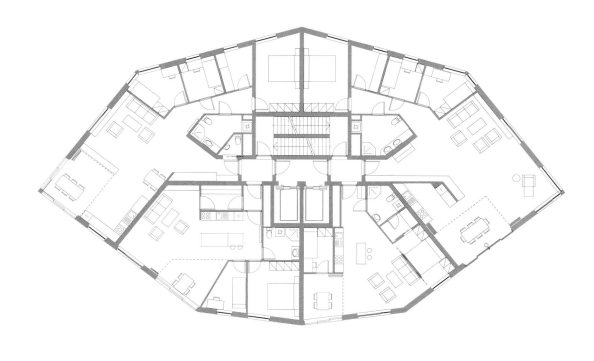

0 1 2 3 4 5   10m

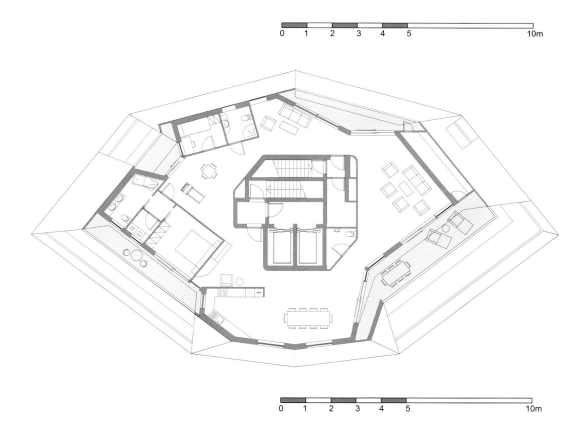

0 1 2 3 4 5   10m

The organizing element of the entire project is the star-shaped core structure of concrete, where the fins are extended as separation walls between the flats. On the ground floor, the fins and bracings of this stem-like core spread out as roots which integrate social meeting places, play and training facilities, generous entrance halls and communal gathering spaces. The orientation and floor plans of the flats are optimized according to views and sun, and integrated winter gardens with folding doors create a flexible quality to the living space.

The three towers have been shaped to allow for diagonal views. The surfaces are divided into a triangular cladding pattern which creates different light shades for each element. The towers have solar collectors on the roofs, as well as geothermal heat pumps in the ground.

项目的整体结构元素是星形混凝土结构，延伸出的两翼作为单位之间的分隔墙。首层主干的延伸结构和支柱向外扩展，作为根部，能容纳社交聚会场所、娱乐和健身设施，以及高大的大厅入口和公共交流区。楼层的布局和朝向都根据阳光的方向和视野进行了优化调整，有折叠门的冬季花园营造出一个灵活、高质量的生活空间。

三栋塔楼之间互成角度，形成了对角线视野。表面被切分成三角形覆层的形式，为每个单位带来了不同的光照效果。大楼的屋顶覆盖着太阳能集热板，而地下则埋藏着地源热泵。

STUDENT HOUSING, PARIS, FRANCE

法国巴黎学生公寓

Architect: MÉTRA+ASSOCIÉS Brigitte Métra
Client: SEMIDEP
Location: 11th District, Paris, France
Site Area: 1,950 m²
Status: Completed
Units: 62
Photography: Philippe Ruault

设计公司：MÉTRA+ASSOCIÉS Brigitte Métra
客户：SEMIDEP
地点：法国巴黎 11 区
占地面积：1 950 m²
状态：已建成
户数：62
摄影：Philippe Ruault

The original construction cost is € 3,790,000
that means: construction cost in China is

¥ 7,265,583

per unit cost is ¥ 117,187

原建造成本为 3 790 000 欧元，
国内建造成本约

¥ 7 265 583 元

国内平均每户造价约 117 187 元。

The site, narrow and elongated, led us to propose a building that stretches along the streets allowing double oriented rooms – with views both on the street and on the new garden. Indeed, the building is naturally and smoothly connected to existing buildings, freeing space for a garden at the heart of the plot and, thanks to open air gangways, offering double-oriented flat. Color is used here to express the vitality of both the area and building tenants. The coloured vibrations of the facades bring life to the street. At the heart of the plot, a microcosm is created. The plants invade the inner space and the facades.

本案场地狭窄细长，设计师因此提出沿街设计建筑物，建成两个朝向的房间——可观看街面和新花园。新建筑物与现有建筑物自然衔接，中心是花园区域、露天过道和双朝向的平台。使用色彩来表达该区域和楼宇的活力。外墙色彩变化也为街道带来生机。中心区域设计了一个小景观。外立面和内部空间的设计加入了植物元素。

Cost and Creativity 成本与创意

The choice of the façades materials has been made for its great creative potential within a tight budget.

立面材料的选择注重其创造潜力，并严格控制在预算之内。

Horizontal and large painted steel blinds cover the building like a "rainbow" echoing the diversity of the neighbourhood, and Parisian facades.

大型水平彩钢百叶窗覆盖着建筑物，形成了一种极富层次感的彩色立面，像彩虹一般表现社区和巴黎外观的多样性。

Close to a lively neighbourhood, the street of the Fontaine au Roi seemed somewhat "abandoned" and deserved to receive from a touch of life, echo of the mixed and animated area. Direct dialogue is established with the surrounding district and the inhabitants of the city. The rounded angle echoes the rounded shape of the opposite building. The skin on the ground floor bends away from the edge of the sidewalk and leads us to the entrance of the residence.

靠近繁华的街区，看似随意"丢弃"的喷泉，聚合生活的点滴感触，映射街区的万象生机，与周边地区和城市的居民建立起直接对话。圆形转角与圆形建筑相对应，外墙贴人行道边弯曲蜿蜒通向居住区。

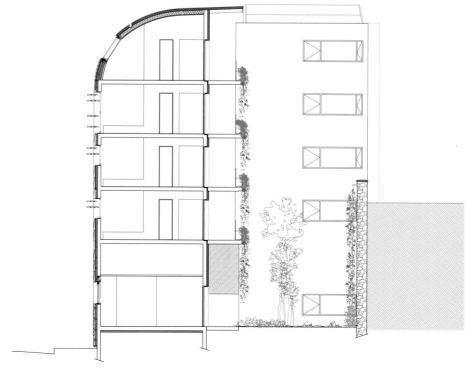

Horizontal and large painted steel blinds cover the building. A variation in four colours, from brick red to orange sunsets, identifies each room. They form a gradation, like a "rainbow" echoing the diversity of the neighbourhood, and Parisian facades. They also are a memory the burgundy color of the former building's windows.

The blades of the blinds, differentiated by the various colours and inclinations, identify each unit and create the vibration of the facade. Expressing the inner life of each student living in the building, the blinds open and close like eyelids on a face. Vibration of color in the streetscape, the student housing belongs to the site bringing a lively and personalized contemporary touch. The facades are moving along with the different hours of the day and night: The occupants, depending on their desire and the direction of their homes, open their own front largely for a view down the street or tilt it to protect from the sun, or close them for the night.

大型水平彩钢百叶窗覆盖着建筑物。色彩有四种变化，从砖红色到日落黄，以区别各个房间。它们形成了一种层次色彩，像彩虹一般表现社区和巴黎社区的多样性。设计中也保留原有建筑物窗户的深紫红色。

百叶窗的叶片，具有多样化的颜色和斜角，以区分各个单元并营造出外墙的变化。为表达建筑物内每个学生的生活，百叶窗打开和关闭就像人的眼帘。外立面色彩随着昼夜不同时段而变化，居住者根据自己的愿望和房屋方向，可打开大部分的门观看街景，或将门倾斜遮挡太阳或关上门度过夜晚。

The space is expanded by the light entering from two sides of the room: south-east and north-west. A 1.10 m wide stripe including space for the bed, bathroom and kitchenette, frees a "space" kind of 2,5 m x 5 m "loft", allowing great flexibility. The sober and minimal atmosphere - white walls and rough concrete, furniture in birch plywood on wheels - will allow each student to arrange the space at will.

空间按房间的光照分两部分：东南部和西北部。一道 1.1 m 宽带状范围内有床、浴室、厨房，附带 2.5 m x 5 m 的阁楼，利用更灵活，冷清狭小的氛围——素墙裸柱、威尔士白桦家具——可让每个学生如愿安排空间。

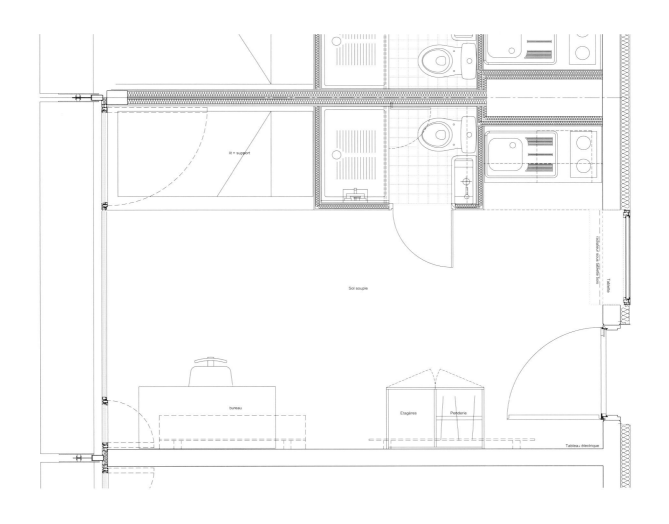

HENIUS HOUSE

HENIUS 公寓

Architect: C. F. Møller Architects
Client: Himmerland Housing Association
Location: Ditlev Bergs Vej, Aalborg, Denmark
Built Area: 16,000 m²
Status: Under Construction
Units: 285

设计公司：C. F. Møller 建筑事务所
客户：Himmerland 住房协会
地点：丹麦奥尔堡 Ditlev Bergs Vej
建筑面积：16 000 m²
状态：建设中
户数：285

The original construction cost is € 35,300,000
that means: construction cost in China is

¥ 64,035,856

per unit cost is ¥ 224,687

原建造成本为 35 300 000 欧元，
国内建造成本约

¥ 64 035 856 元

国内平均每户造价约 224 687 元。

Aalborg has a great deal of youth housing, scattered throughout the city. 240 new youth residences and 45 family homes – called Henius House after IsidorHenius, who owned the Eternit cement factory site and laid the foundation for the industrial activities there – represent a proposal for a centralised, modern youth housing project that can offer something more, including a strong social community with many opportunities for common activities.

奥尔堡拥有大量的青年住房，散落在整个城市。本案总共包括 240 套青年公寓和 45 套家庭公寓——以 IsidorHenius 的姓氏为名，称作 Henius 公寓。IsidorHenius 拥有石棉水泥厂场地，并在那里打下了工业活动基础。本案的设计呈现了一种集中化的现代新青年住宅，为人们提供了更多的东西，包括可以创造共同活动机会的超大社区。

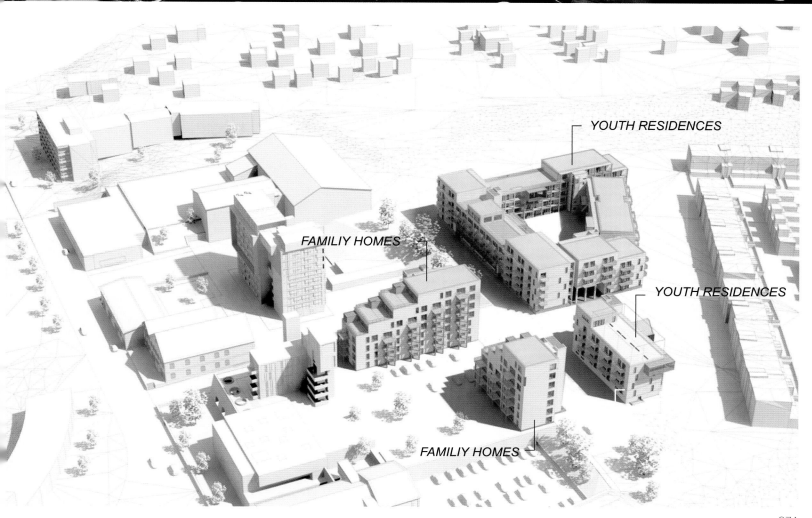

YOUTH RESIDENCES

FAMILIY HOMES

YOUTH RESIDENCES

FAMILIY HOMES

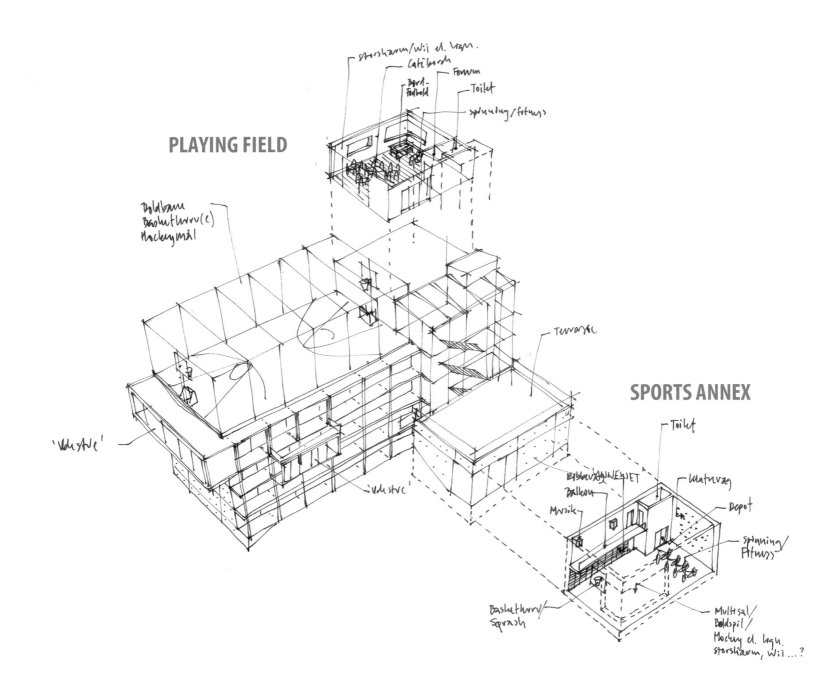

PLAYING FIELD

SPORTS ANNEX

Simple and cost-efficient pre-fabricated building components are used to deliver a maximum of area for a minimum of cost. Generous and attractive shared facilities provide a high living standard at a low expense.

预制建筑构件生产简单且造价低廉，可以用最低的成本建造最大面积。宽敞且便利的公共设施在低费用的前提下为人们提供更高的生活水准。

Various colourful cubes set in volumes brightly and vividly.

镶嵌在建筑体量上的彩色结构体明亮而生动。

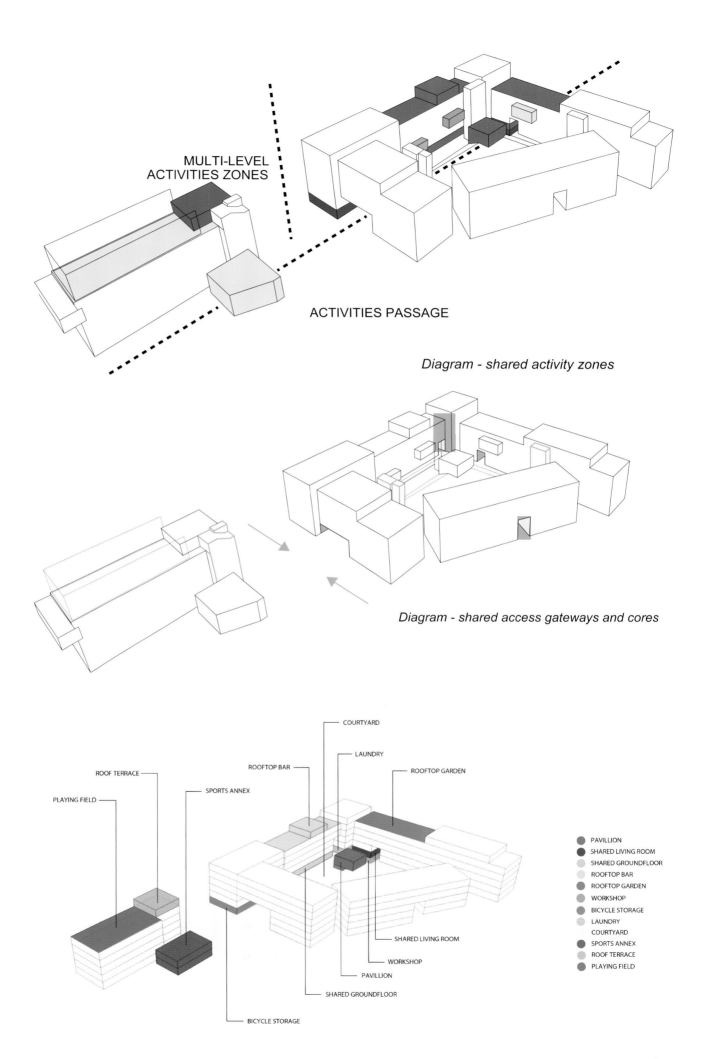

MULTI-LEVEL
ACTIVITIES ZONES

ACTIVITIES PASSAGE

Diagram - shared activity zones

Diagram - shared access gateways and cores

COURTYARD

LAUNDRY

ROOFTOP BAR

ROOFTOP GARDEN

ROOF TERRACE

SPORTS ANNEX

PLAYING FIELD

SHARED LIVING ROOM

WORKSHOP

PAVILLION

SHARED GROUNDFLOOR

BICYCLE STORAGE

PAVILLION
SHARED LIVING ROOM
SHARED GROUNDFLOOR
ROOFTOP BAR
ROOFTOP GARDEN
WORKSHOP
BICYCLE STORAGE
LAUNDRY
COURTYARD
SPORTS ANNEX
ROOF TERRACE
PLAYING FIELD

The youth residences will have facades of grey concrete with exposed construction joints, inspired by the cement factory's original rough industrial architecture on the site. The rough concrete panels are interspersed with protruding volumes, another characteristic of the former industrial buildings. These volumes contain the common facilities, which are indicated in various bright colours.

青年公寓的外立面受水泥厂原工厂建筑的启发，裸露灰色混凝土和结构缝。粗糙的混凝墙板中凸出各种颜色鲜艳的结构体，这些结构体包含公共设施，是以前工业建筑物的另一个特点。

Six buildings in all, with four to five storeys each, will be oriented so that all residents are secured the best possible light and views of the area – which is on a ridge overlooking the city, the valley of Østerådal and the Limfjord strait.

In addition to common kitchens and common rooms on the various floors, there will also be common fitness facilities, an Internet café and workshops, and outdoor areas suitable for social activities. The roof areas can for example be used for ball games or Friday parties.

The youth residences are flexible in their design to ensure optimal opportunities for future adaptations of the interiors, so that two youth residences for example can be merged into a family home or a shared flat. Sustainability features include low-energy windows and highly insulated building envelopes, solar panels, ventilation with heat recovery, green roofs and rainwater harvesting.

整个项目一共六栋建筑物，每栋四到五层，其朝向设计使所有居民都能获得最好的光线和视野，越过屋脊，远眺城市、山谷和海峡。

各楼层除了公用厨房和休息室，也将安装公用健身设施，设立网吧和工作间，以及适用于社交活动的区域。屋顶可举行像球类或周末舞会类活动。

青年住宅的设计灵活，以适应未来的使用，例如两个青年住宅可以合并成一个家庭或一个共享单位。可持续发展的功能包括节能窗和高度隔热的建筑围护结构、太阳能电池板、带热量回收的通风设施、屋顶绿化和雨水收集。

SUNFLOWERS

向日葵公寓

Architect: Carlos Arroyo Arquitectos
Client: Mechelse Goedkope Woning / VMSW
Location: Mechelen, Belgium
Units: 400

设计公司：Carlos Arroyo Arquitectos
客户：Mechelse Goedkope Woning / VMSW
地点：比利时 Mechelen
户数：400

We call them sunflower buildings. They come in different "sizes". They are arranged in clusters, following these principles:

Maximize the use of south oriented areas, both inside and outside the buildings. All day rooms are oriented to the south, with larger windows and larger spaces. On the ground floor, usable garden areas are also to the south of buildings. Open areas are protected from vehicle noise in main road, being at a higher level and screened of by the parking shelters. The section of all the constructions is designed for minimum shadow, both on other buildings or on open space.

本案被称为向日葵公寓。它们大小不同，并且以群组的形式组合在一起，遵循以下的原则：

建筑内外都最大化利用场地朝南的空间，日用房都朝南，设有大型窗户和开阔空间。首层的花园也设置在建筑的南部。公共区设置在高处，以屏蔽公路上的车辆噪声，并设有停车棚。各栋公寓在设计上考虑到了阳光被遮挡的可能性，使每栋楼和公共区都能接受最多的阳光。

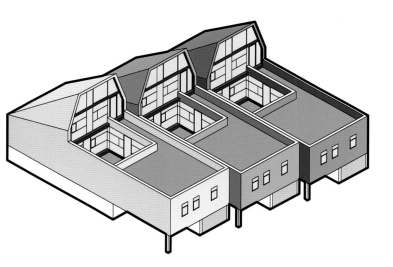

In the existing situation in Mechelen, public space is an expanse of no-man's land in sharp contrast with the privacy of each flat. There is no interaction space.

In our proposal, there is a gradient from public to public. There is such a gradient in more traditional community situations, but in our proposal this is redesigned, to match the needs of present-day communities. Our design method includes a series of workshops with the local community, to establish the most adequate deal. Our aim is to redefine shared space.

There is interaction space at all levels of the proposal: in open public space, inside the clusters, in each building, and within the flats.

Mechelen 的现状是公共空间开阔无人，与每栋公寓的私人空间形成强烈对比，两者间没有互动的空间。

设计师的提案是设计一个公共空间之间的过渡。虽然在以往的传统社区里也有类似的过渡，但在本案中，设计师重新进行了设计，以符合现今社区的要求。其方法包括在社区中设立一系列的工作坊，以建立最适当的过渡空间。设计师的目标是重新定义共享的空间。

提案中还对公寓的各层设置了互动空间，包括开放的公共空间、公寓楼群间、每栋公寓楼间和每间公寓房间。

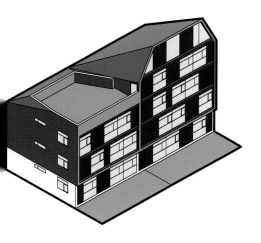

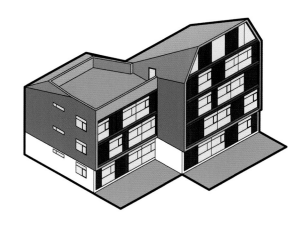

Cost and Creativity 成本与创意

The key to success in this project is to provide the basis for a new social network. Lots of shared spaces help save places while providing a social community.

本案成功的关键在于提供了一个新的社会系统基础。大量的共享空间有助于节省土地资源，又能为住户提供一个便于交流的社区场所。

To ensure a sense of belonging each construction is distinct, and is finished in a different variety of the same material.

为了让公寓有归属感，每栋公寓都各有特色，并且是使用相同的材料来达到多种效果。

Within the flats, there is a small undefined space that can become a playing area for children or a place for a tele-worker to have a small office.

In some of the buildings there are rooms that the neighbours can use in different ways. Someone may use it for two hour a day and offer tai-chi lessons.

In the buildings for the elderly (or for singles/couples), the shared space can be more domestic, with a shared lounge, a solarium, a laundry, a HBO, or a kitchen and dining area, where the more sociable residents can optimise their efforts, as well as establishing a social network.

Some of the shared spaces have a broader scope, such as the children's room, where residents in the building (and maybe in

the neighbourhood) can organise shifts to look after each other's children.

Bicycle storage and parking areas are already familiar shared spaces.

Playgrounds on the south side of the buildings, and places to sit around, provide the opportunity to make the most of any sunny day that luck may bring, and a chance to meet your neighbours.

Each construction includes a specific shared space. Shared space is always in the best place of the building, to make sure that people will want to use it (as a back-up, if a specific shared space is not successful, these spaces will be valuable for other uses, or even turn into excellent flats).

每间房内都有未确定用途的小空间，可以作为儿童游乐区，或者作为远程工作者的小型办公室。

部分公寓楼内安排了空房供住户使用，可供个人每天使用 2 小时，或者作为太极拳教室。

在老人（或单身／家庭）公寓里的共享空间也可作私人用途，里面有共用的休息室、日光浴室、洗衣房、家庭影院、厨房和用餐区。喜欢交际的居民可以充分利用这里，建立更好的邻里关系。

部分共享空间拥有更开阔的空间，如儿童房，居民在室内或附近的社区可以轮流照顾孩子们。

自行车和汽车停放区也是大家熟悉的共享空间。

操场设置在公寓的南侧，让住户可以围坐一起交流，享受日光，联络感情。

每栋楼内都有一个特定的共享空间，通常位于楼内最好的方位，以鼓励用户多加交流。该空间还有后备用途，如果使用率较低，这里则会做其他用途，甚至可以改造为豪华公寓套房。

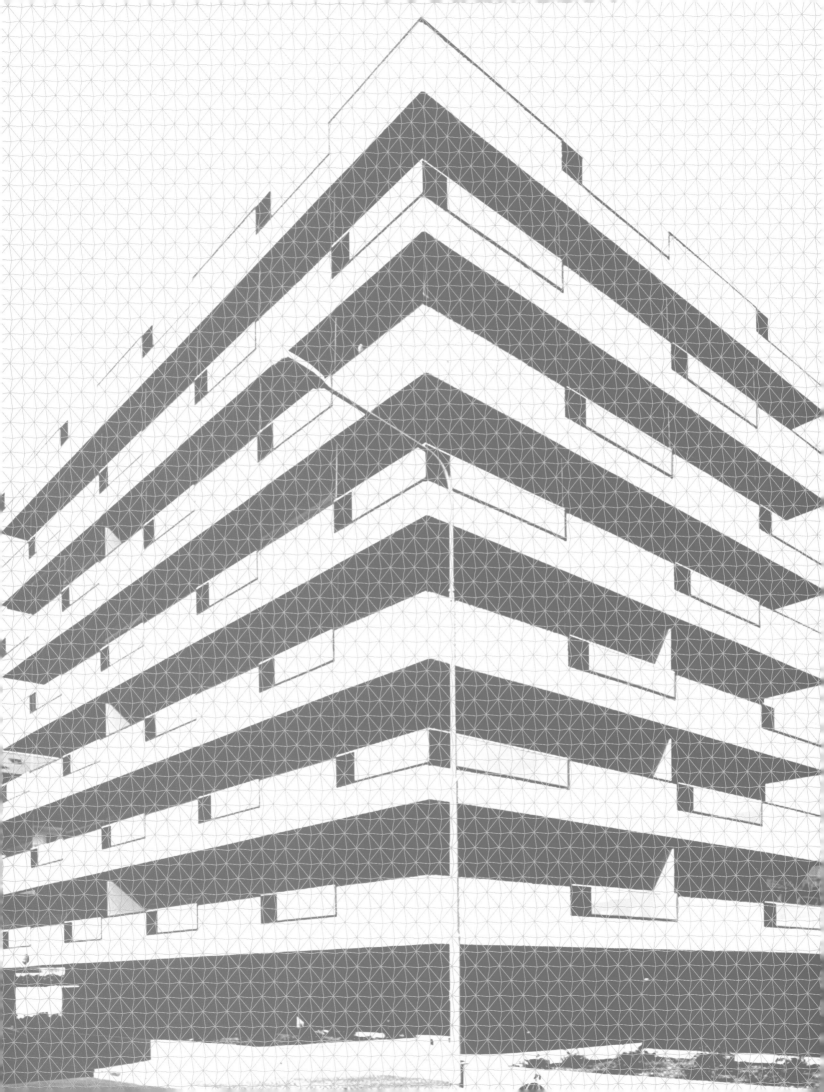

ENERGY

节能

BASTYR UNIVERSITY STUDENT HOUSING VILLAGE

BASTYR 大学学生公寓

Architect: CollinsWoerman
Location: Kenmore, WA, U.S.A.
Site Area: 23,472 m²
Status: Completed
Units: 132

设计公司：CollinsWoerman
地点：美国华盛顿州肯莫尔
占地面积：23 472 m²
状态：已建成
户数：132

The original construction cost is $ 12,500,000
that means: construction cost in China is

¥ 13,747,402

per unit cost is ¥ 104,147

原建造成本为 12 500 000 美元，
国内建造成本约

¥ 13 747 402 元

国内平均每户造价约 104 147 元。

As the goal was to provide a sustainable living environment enabling students to become "future leaders in natural health and sciences that integrate mind, body, spirit and nature," the design team conducted a series of outreach sessions and design charrettes with the client, students and community to develop an overarching theme of "connection" — to the land, to the community and to each other. Sustainability goals, such as reducing the need for cars, water, and energy, and strategies for site and building design, such as optimal solar orientation, were implemented. The team ultimately arrived at a solution that breaks away from traditional dormitory or apartment-style buildings, thereby establishing a new typology for student housing: one that connects more intimately with the ground and allows greater interaction among students.

公寓的目标是提供一个可持续发展的生活环境，让学生在"天然健康和科学的环境下整合头脑、身体、精神和自然，成为未来的领导人"。设计团队在经过一系列的研讨，并听取了专家组、客户、学生群体和社区的意见后，决定采取"联系"的设计主题——将土地、社区和公寓各部分彼此联系起来。公寓彻底贯彻可持续发展的目标，如减少对车辆、水和能源的需求，建筑的设计也考虑了获取日光的最佳方向。最终设计团队成功打破了传统宿舍和公寓形式的模式，创造出新式的学生公寓，与自然接触更多，鼓励学生之间的交流互动。

Cost and Creativity 成本与创意

Designing smaller units and eliminating unnecessary space helped achieve significant cost savings and to create new opportunities for greater energy efficiency.

设计注重建造更小型的单位，并消除不必要的空间，这样有助于实现成本控制，并为提高能源效率创造新的机会。

The project embodies collaboration, resource conservation, and dedication to natural health and well-being.

该项目体现了优化配置、节约资源的特性，为保证建筑的天然健康和环保服务。

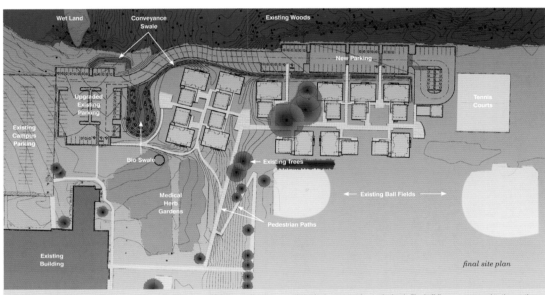

final site plan

Storm water run-off from parking and roadway surfaces is directed into a conveyance bioswale and then into a biofiltration swale. Final treatment occurs in a leaf compost media filter. The treated run-off is held in a 293,000 gallon underground storm water detention tank for eventual release to existing wetlands to the north of the site. As the design evolved, pervious area was reduced and a decision was made to keep four large evergreen trees on the plateau and place the buildings around them, creating a further connection to the land. The buildings were moved to the south to allow parking to miss the trees. Paved areas at each cottage will be sand set pavers in lieu of concrete. The parking areas and roadway were originally planned as pervious concrete; but that site area too was needed to recharge wetlands and the proximity to the evergreen trees meant that the fine needles would eventually clog the open voids and render the pervious paving less useful.

The architect designed eleven, 4,600-sf, three-story cottages housing 12 students each. Each cottage contains private bedroom/bathroom combinations with communal kitchen, great room, and study area. The village is the USGBC's first LEED for Homes™ Platinum-certified student housing project on the West Coast. In 2010, the Student Housing Village was awarded USGBC's Project of the Year for "Outstanding Multifamily Project."

公寓楼共有 11 栋，高 3 层，每座占地约 427.35 m²，可住 12 名学生。每栋公寓都有私人房间和浴室、公用厨房、大厅和学习区。本项目是美国绿色建筑委员会（USGBC）中首个获得白金认证的西海岸住房项目，并在 2010 年被授予"杰出综合住宅"奖。

The campus is situated in a cleared area of a 52-acre existing conifer forest in Kenmore, Washington. The project's original location would have required significant tree removal and possible encroachment onto an existing wetland, and 180 required parking stalls would have been displaced. The design team successfully advocated for a different location that not only saved existing parking, but allowed for a more protected site that would require less infrastructure for future phases.

校园位于华盛顿肯莫尔区，坐落在一片占地 210 436.72 m^2 的针叶林空地上。项目原本的位置需要移走大量的树木，并侵占到原来的湿地，有 180 个车位的停车场也会被移除。设计团队成功地选择了另一个位置，不仅可以保存停车场，对现有设施和未来设施的要求也相应较少，以保护环境。

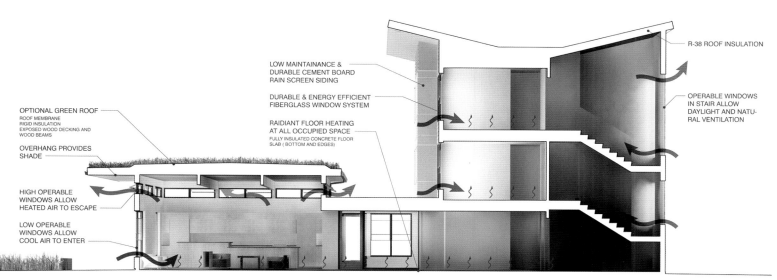

OPTIONAL GREEN ROOF
ROOF MEMBRANE
RIGID INSULATION
EXPOSED WOOD DECKING AND
WOOD BEAMS

OVERHANG PROVIDES
SHADE

HIGH OPERABLE
WINDOWS ALLOW
HEATED AIR TO ESCAPE

LOW OPERABLE
WINDOWS ALLOW
COOL AIR TO ENTER

LOW MAINTAINANCE &
DURABLE CEMENT BOARD
RAIN SCREEN SIDING

DURABLE & ENERGY EFFICIENT
FIBERGLASS WINDOW SYSTEM

RAIDIANT FLOOR HEATING
AT ALL OCCUPIED SPACE
FULLY INSULATED CONCRETE FLOOR
SLAB (BOTTOM AND EDGES)

R-38 ROOF INSULATION

OPERABLE WINDOWS
IN STAIR ALLOW
DAYLIGHT AND NATU-
RAL VENTILATION

DAYLIGHTING, VENTILATION, ACOUSTICS AND ENERGY EFFICIENCY DIAGRAM

The cottages are designed to allow a balance between privacy and connection to other students. The orientation of the great room spaces allows the residents of each cottage to have a visual connection with students in the other cottages, and to the landscape and exterior gathering spaces, enhancing the sense of community.

Buildings and pathways were intentionally sited to create a strong sense of community and collaboration. High albedo concrete walks connect the cottages to each other and the rest of the campus, and over 30 built-in concrete benches

encourage outdoor living throughout the Student Village. Each cottage has an attached secure, covered bike storage area with sliding doors and hanging racks.

The Great Room takes advantage of daylighting to reduce the demand for artificial lighting. Ventilation provided by the operable windows and air trickle vents is circulated throughout the room by the ENERGY STAR-rated ceiling fan.

The wall and door between the study and the hallway is tempered glass to transmit light. Sand-set pavers are used at exterior decks off the study areas of each cottage.

公寓的设计在开放和私密间保持微妙的平衡。每栋楼通过大厅都可以与其他公寓楼、环境、户外交流场所有视觉连接，加强社区的整体感。

各栋公寓和走道的位置都经过特意设置，以加强社区内的联系和增加交流机会。具有高反照率的混凝土走道连接各栋公寓和校园，30 多张户外长椅为学生的户外活动提供场所。每栋公寓都配置了有顶棚、安全性高的自行车存放区，通过滑动门进入，并提供有挂衣架。

大厅的采光良好，可减少人工照明的需求。可手动打开的窗口、通风口和天花板的节能风扇保证大厅的空气流通。

书房和走廊之间的墙和门都是钢化玻璃，让光线能照射进来。每栋公寓的学习区前的空地都采用沙面铺装。

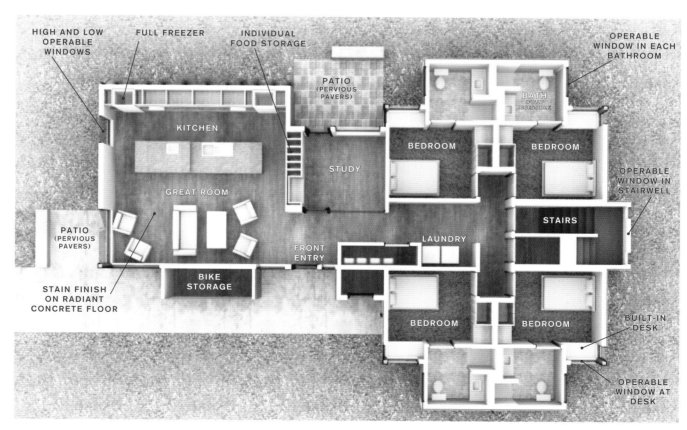

HIGH AND LOW
OPERABLE
WINDOWS

FULL FREEZER

INDIVIDUAL
FOOD STORAGE

OPERABLE
WINDOW IN EACH
BATHROOM

PATIO
(PERVIOUS PAVERS)

BATH
FULLY ACCESSIBLE

KITCHEN

BEDROOM

BEDROOM

STUDY

GREAT ROOM

OPERABLE
WINDOW IN
STAIRWELL

PATIO
(PERVIOUS
PAVERS)

LAUNDRY

STAIRS

FRONT
ENTRY

STAIN FINISH
ON RADIANT
CONCRETE FLOOR

BIKE
STORAGE

BEDROOM

BEDROOM

BUILT-IN
DESK

OPERABLE
WINDOW AT
DESK

ARMSTRONG SENIOR HOUSING

阿姆斯特朗老年公寓

Architect: David Baker + Partners Architects
Associate Architect: Full-Circle Design Group
Client: BRIDGE Housing
Location: San Francisco, California, U.S.A.
Site Area: 35,000 m²
Status: Completed
Units: 116
Photography: Brian Rose

设计公司：大卫·贝克及合伙人建筑事务所
合作公司：Full-Circle 设计集团
客户：BRIDGE Housing
地点：美国加州旧金山
占地面积：35 000 m²
状态：已建成
户数：116
摄影：布赖恩·罗斯

The original construction cost is $ 41,200,000
that means: construction cost in China is

¥ 45,311,436

per unit cost is ¥ 390,616

原建造成本为 41 200 000 美元，
国内建造成本约

¥45 311 436 元

国内平均每户造价约 390 616 元。

Located in San Francisco's Bayview District, this affordable senior housing complex is served by a nearby stop of the Third Street Rail, the new Muni line. These four levels of senior housing above neighborhood-serving retail are part of a larger trend of transit-oriented development along this corridor, and will catalyze future such projects, bringing density, variety and services to the area. Armstrong Senior Housing is a HUD 202 project.

公寓位于旧金山贝维尤区，毗邻第 3 大道的新 Muni 线铁路站。高 4 层，底层是社区商场，公寓代表了该区以交通为导向的住宅项目发展趋势，并将促进更多此类型项目的建设，为该区带来更多的居民、多样性以及丰富的服务设施。本项目是美国住房和城市发展部（HUD）的 202 号项目。

To reflect the historically African-American population of the neighborhood, the color palette is drawn from traditional African textiles—the deep indigos and bright accents of Ghanian dutch wax resist fabrics—which along with the window placement, appears to wrap the public face of the building in an interlocking "quilt" of color and pattern. The private side is cloaked in the more subdued tones drawn from the earthy hues of Malian mudcloth.

为了能反映出该区历史上的非洲裔人口，公寓的色调选自非洲的传统纺织品——深靛蓝与非洲加纳和荷兰的抗蜡面料中明亮的艳色，和窗口一起形成奇妙的组合，看上去就像是一张色块和条纹拼接而成的"被子"，将公寓向外的立面紧紧包裹起来。公寓向内的立面则用较为柔和的色调，选自马里泥色布料的土色系。

Parking is reduced to realistically reflect the auto-ownership of the population and capture additional square footage for retail services. There is a car-share pod with two available vehicles as well as secure bicycle parking. The retail space features a dedicated shower and changing area, to facilitate bicycle commuting.

公寓特意减少停车位，以真实反映出该区人均汽车拥有量，并将多出的空间作为商场的额外零售区域。其中的一个小型停车场可以容纳两辆汽车，其他空间作为自行车停放区。商场还设有专门的沐浴和更衣区，以方便骑自行车出行的居民。

Cost and Creativity 成本与创意

The project combines a variety of convenience facilities with rich green, to meet the needs of future urban development.

公寓结合了各种便民设施，绿化丰富，可以满足未来城市发展的需求。

This project is being designed and built to a LEED Gold standard, with healthy interiors for senior residents. Photovoltaic arrays will provide solar electric power and domestic hot water.

本项目的设计和建设目标都以 LEED 金奖为标准，旨在为老年居民提供和谐健康的室内居所。其光电池设备为公寓提供太阳能发电和热水。

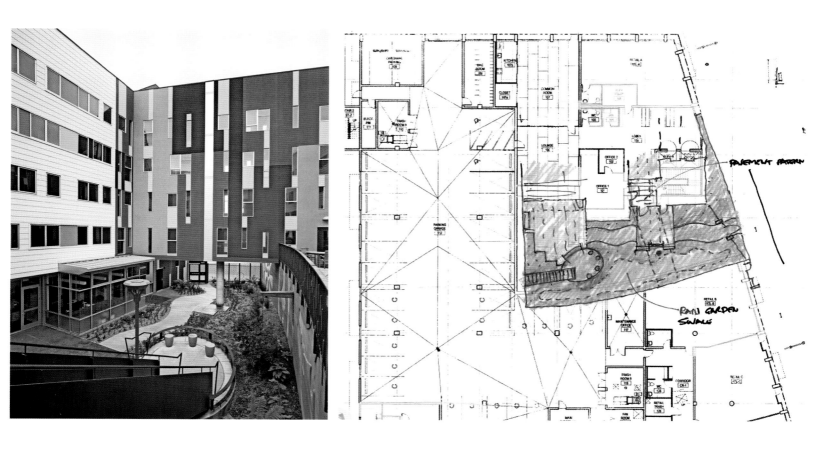

The residences—predominantly studios and one-bedroom units—enclose a courtyard and sit atop commercial space set to house shops, senior services, a library, and a community center. The unique landscaping—vegetated bioswales along the street and mews, and a courtyard rain garden—does double-duty, adding green areas and creating miniature wetlands that manages runoff, easing the burden on the city's combined stormwater and sewage system.

公寓的户型以工作室和单卧套房为主，配有小庭院，楼下的商业空间设备齐全，有零售店、高级服务场所、图书馆和社区中心等。沿着街道和小道分布的庭院植物葱郁茂盛，别具一格。庭院同时也作为雨水花园，具有双重功效：既增加了绿化率，又创造了一片微型湿地，有效管理雨水径流，结合城市的雨水和污水处理系统，为城市减轻负担。

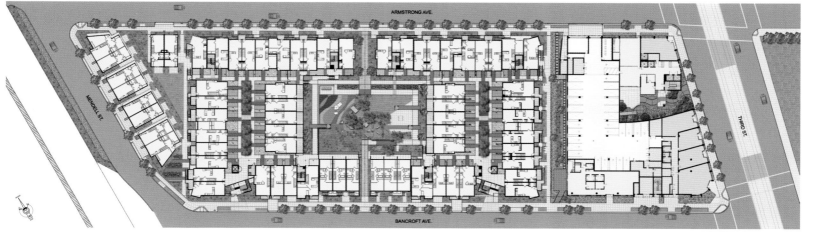

ZUIDERZEEWEG STUDENT HOUSING

ZUIDERZEEWEG 学生公寓

Architect: Fact Architects
Client: De Key
Location: Zuiderzeeweg, Amsterdam, The Netherlands
Site Area: 12,395 m²
Status: Completed
Units: 335
Photography: Luuk Kramer

设计公司：Fact 建筑事务所
客户：De Key
地点：荷兰阿姆斯特丹 Zuiderzeeweg
占地面积：12 395 m²
状态：已建成
户数：335
摄影：Luuk Kramer

The original construction cost is € 11,500,000
that means: construction cost in China is

¥ 18,373,466

per unit cost is ¥ 54,846

原建造成本为 11 500 000 欧元，
国内建造成本约

¥ 18 373 466 元

国内平均每户造价约 54 846 元。

Zuiderzeeweg is a housing project for students with 335 apartments.

The location is temporary, in the future all the buildings will move to a definitive location.

The building blocks are made from high quality units. The units are prefabricated with balcony, windows, floor heating system and sanitary. After completion the units were transported to the location.

Zuiderzeeweg 学生公寓将为学生提供 335 套住房。

公寓目前的选址还只是暂定，在日后选址确定后所有的建筑单位才搬到那里。

建筑由各个高品质的单位组成。各单位都是预先制定，有阳台、门窗、地板采暖系统和清洁设施。单位建成后将被运到项目所在的地址。

Cost and Creativity 成本与创意

The open collaboration resulted in a project built within budget and on time. The buildings are movable, sustainable and prepared for future environmental requirements.

开放的合作是项目在预算范围内得以建成的前提。这栋可移动的公寓不仅符合可持续发展要求，还考虑到了将来的环保要求。

The buildings are sustainable and prepared for future environmental requirements.

公寓贯彻可持续发展理念，可满足将来环境的要求标准。

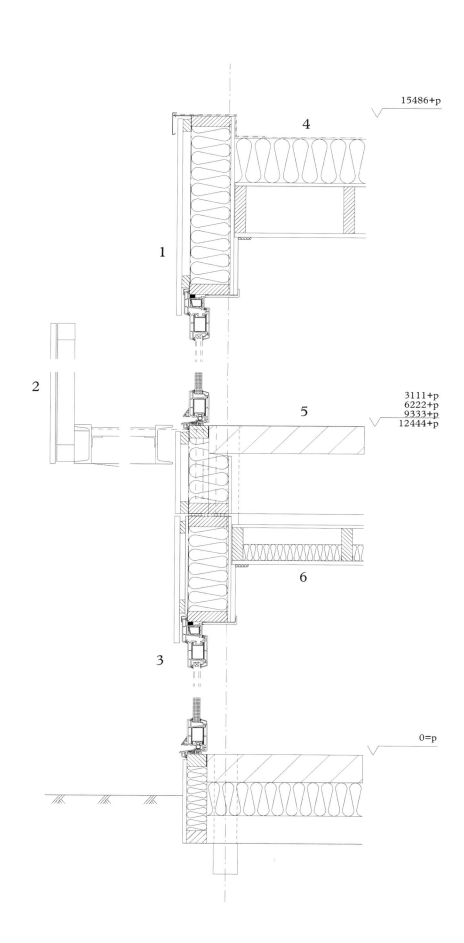

15486+p

4

1

2

3111+p
6222+p
9333+p
12444+p

5

6

3

0=p

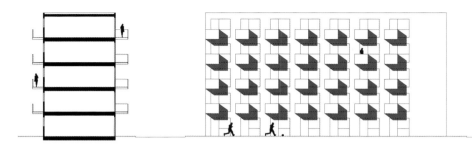

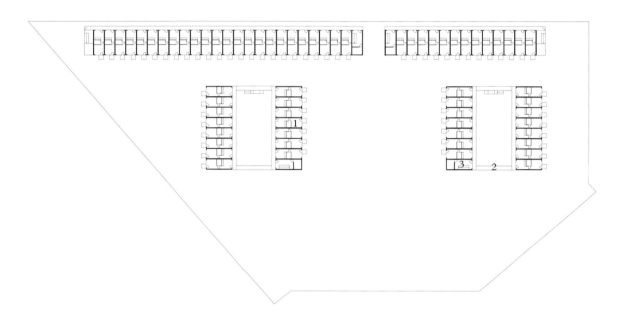

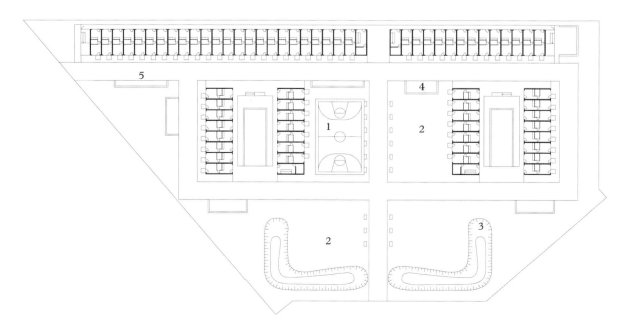

Fact Architects designed a durable and robust facade. The dark wooden facade with the orange texts give the complex a sustainable and strong appearance. The exterior area is designed with the same clear approach. The buildings and grass hills enclose a courtyard consisting of lawns and an orange sports field. Along the ring road bicycle parking bays and spaces for bins are situated, separated by with roses overgrown walls. At the main road concrete blocks are placed that can be used as seating.

立面的材料坚固且耐用。暗色的木质立面配上橙色的样式赋予整个公寓极具冲击力的抢眼外观。公寓和长满青草的山丘一起围合出一座庭院，里面有草坪和橙色的运动场。沿着环形公路设置了自行车停车位和垃圾箱，以在墙上茂密丛生的玫瑰作为间隔。设置在主要道路上的混凝土块可作为座椅使用。

MNM HOUSING

MNM 住宅

Architect: GEZA
Client: Min.I.Mal. srl, Immobiliare Marghe
Location: Udine, Italy
Site Area: 1,400 m²
Status: Completed
Units: 19
Photography: Alta Risoluzione

设计公司：GEZA
客户：Min.I.Mal. srl、Immobiliare Marghe
地点：意大利乌迪内
占地面积：1 400 m²
状态：已建成
户数：19
摄影：Alta Risoluzione

The original construction cost is € 3,500,000
that means: construction cost in China is

¥ 7,743,046

per unit cost is ¥ 407,529

原建造成本为 3 500 000 欧元，
国内建造成本约

¥ 7 743 046 元

国内平均每户造价约 407 529 元。

Found in a residential area now under development, the seven-storey building is rigorously designed and features four identical modular facades to confer unity and order to a chaotic and unarticulated neighbourhood. All fronts, with balconies from ground to top floor, make a continuous filter between indoor and outdoor space.

本案位于一个发展中的居民小区内，七层楼高，其设计风格严谨，四面外观统一的外墙，在凌乱不清的邻楼群中尤为整齐划一。从正面看，自地面到楼顶，层层阳台连续整齐，内外空间更加通透。

Cost and Creativity 成本与创意

The building presents a double skin: an exterior one, which is more urban and rough, made of concrete and glass; and an interior one, which is more domestic and intimate, which is made of wood.

本住宅有双重表皮：外层由混凝土和玻璃构成，粗犷且有城市感；内层是木制的，显得更为私人和亲密。

Latest energy-saving technology was used to achieve sustainable development in a long period.

项目采用最新的节能技术，以确保长期的可持续发展。

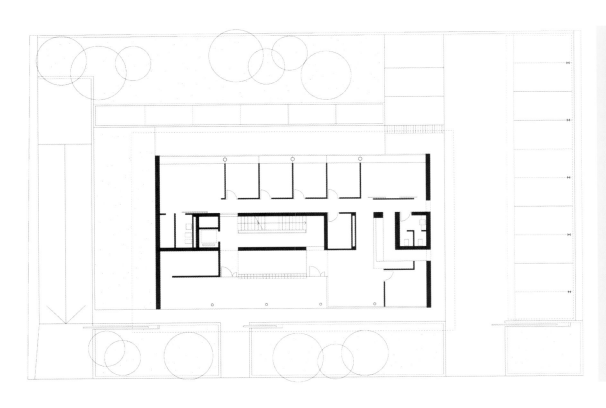

The same rigour applies to the plan of the standard floor where, because the staircase is located asymmetrically at the centre, the different types of flats get totally flexible and prone to be assembled together without needing to alter the modularity of installations and structures.

On the other hand, the connection to the ground of the see-through glazed office floor creates different relations with the site plan, and with pedestrian, cycle and vehicle access ways.

地面也按同样严格的标准设计，因为楼梯不规则地安排在中心。不同类型的公寓灵活搭配在一起而无需改变装置和结构。

而与透明玻璃办公室地板的连接，给平面布置、行人、自行车和汽车出入通道增添了不少变数。

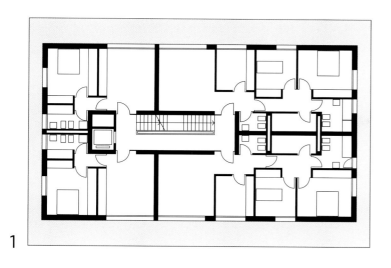

1

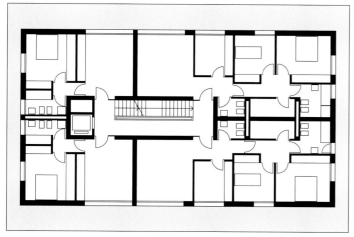

2

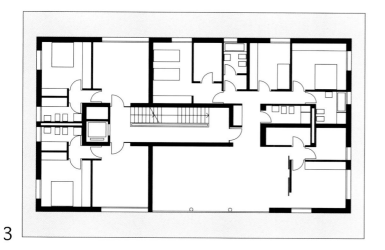

3

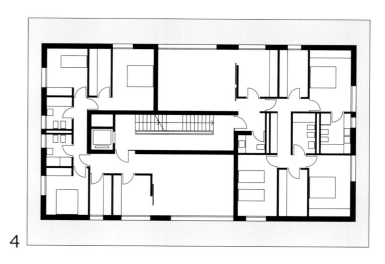

4

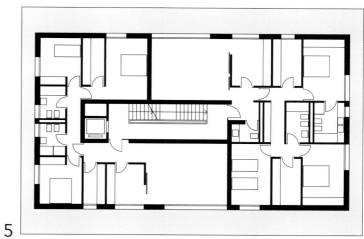

5

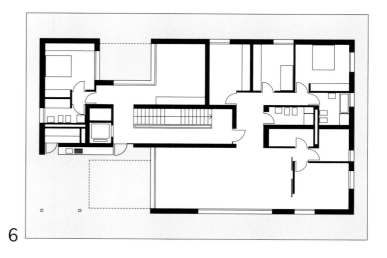

6

Heating and power installations are partly centralized and sized so as to maximize efficiency. To this aim, the latest energy-saving technology was used, in compliance with the forthcoming standards on economy of resources.

供热供电装置统一集中以利发挥最大能效，最新的节能技术也被采用以适应未来的节能标准。

STEP UP ON 5TH

第 5 街 STEP UP 住宅

Architect: Brooks + Scarpa
Client: Step Up
Location: Downtown Santa Monica, California, U.S.A.
Site Area: 2,935.74 m²
Status: Completed
Units: 46
Photography: John Edward Linden

设计公司：Brooks + Scarpa
客户：Step Up
地点：美国加州圣莫妮卡市区
占地面积：2 935.74 m²
状态：已建成
户数：46
摄影：John Edward Linden

The original construction cost is $ 11,400,000
that means: construction cost in China is

¥ 12,537,630

per unit cost is ¥ 272,557

原建造成本为 11 400 000 美元，
国内建造成本约

¥ 12 537 630 元

国内平均每户造价约 272 557 元。

Step Up on 5th is a bright new spot in downtown Santa Monica. The new building provides a home, support services and rehabilitation for the homeless and mentally disabled population. The new structure provides 46 studio apartments of permanent affordable housing. The project also includes ground level commercial/retail space and subterranean parking. The density of the project is 258 dwelling units/acre, which exceeds the average density of Manhattan, NY (2000 USA Census Bureau Data) by more than 10%.

第 5 街 Step Up 住宅项目是圣莫尼卡中心区的新亮点。新建筑物为无家可归者和弱智人士提供了一个家庭、辅助服务和康复服务。新的结构提供了 46 个经济适用型单身公寓。该项目还包括地面商业区 / 零售商场和地下停车场。该项目的密度是 258 住宅单位 /4046 m²，超过纽约曼哈顿的平均密度（2000 年美国人口普查局的数据）10% 以上。

1548

step up

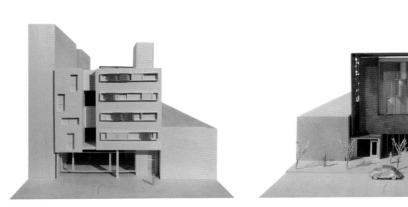

East Elevation

West Elevation

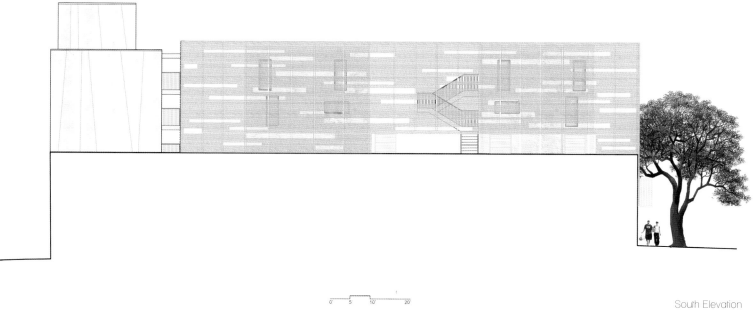

South Elevation

Cost and Creativity 成本与创意

The panels are functional as well as aesthetic, creating shade and privacy, and thus reducing cooling loads and reducing the usage cost after completion.

功能丰富又美观的立面既能起到遮阳作用，又能保护隐私，同时降低制冷负荷，进而降低了建成后的使用成本。

The building is loaded with energy-saving and environmentally benign or "sustainable" devices.

项目建筑是节能环保或"可持续"型设施。

Section A

step up

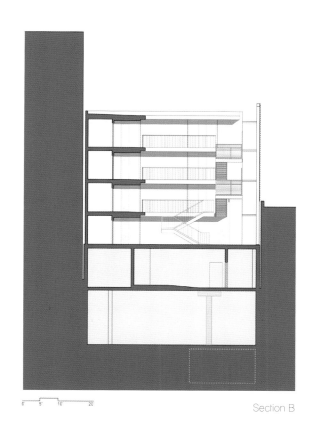

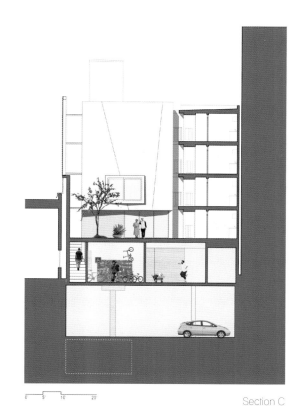

Section B

Section C

Step Up on 5th distinguishes itself from most conventionally developed projects in that it incorporates energy efficient measures that exceed standard practice, optimize building performance, and ensure reduced energy use during all phases of construction and occupancy. The planning and design of Step Up on 5th emerged from close consideration and employment of passive design strategies. These strategies include: locating and orienting the building to control solar cooling loads; shaping and orienting the building for exposure to prevailing winds; shaping the building to induce buoyancy for natural ventilation; designing windows to maximize day lighting; shading south facing windows and minimizing west-facing glazing; designing windows to maximize natural ventilation; utilizing low flow fixtures and storm water management; shaping and planning the interior to enhance daylight and natural air flow distribution. These passive strategies alone make this building 50% more efficient than a conventionally designed structure.

第 5 街 Step Up 住宅项目与最传统的开发项目的区别在于它采用超越常规标准的节能措施，优化建筑性能，并确保在建设和入住的各个阶段减少能源使用。项目的缜密规划设计策略，其优势得到凸显。这些策略包括：对建筑物进行合适的定位和定向，以控制太阳能制冷负荷；合适的建筑物外形和朝向以适应主风向，提高自然通风效率；窗口设计有利于最大限度地自然采光，遮蔽南向窗口，缩小西向窗口；窗口设计利于最大化自然通风；利用低流量装置和雨水管理，加强建筑物内部的自然采光和通风。这些运用使它比传统结构设计的建筑节能效率提高 50%。

Materials conservation and recycling were employed during construction by requiring waste to be hauled to a transfer station for recycling. The overall project achieved a 71% recycling rate. Specifying carpet, insulation and concrete with a recycled content, and utilizing all-natural linoleum flooring also emphasized resource conservation. The project uses compact fluorescent lighting throughout the building and double-pane windows that have a low-E coating. Each apartment is equipped with water-saving low flow toilets and a high-efficiency hydronic system for heat. While California has the most stringent energy efficient requirements in the United States, Step Up incorporates numerous sustainable features that exceeded state mandated Title 24 energy measures by 26%. Although not submitted at this time, the project has followed the LEED certification process and would receive 39 points making it equivalent to LEED "Gold".

项目建设期间，使用环保再生材料，剩余废料必须运到回收站以便再生利用。全项目材料循环利用率达到70%，特别在地毯、绝缘材料和混凝土方面，强调可再生观念，运用全天然地板材料，强调环境保护。整个建筑都采用简单的日光灯照明，双层玻璃窗户降低了辐射，每个公寓都配备节水型洁具和高效循环供热系统，在美国加利福尼亚州有着严格的节能要求，而项目的能源利用比24号强制令提高了26%，虽然没有申报，该项目按照LEED的规程，将获得39项金奖提名。

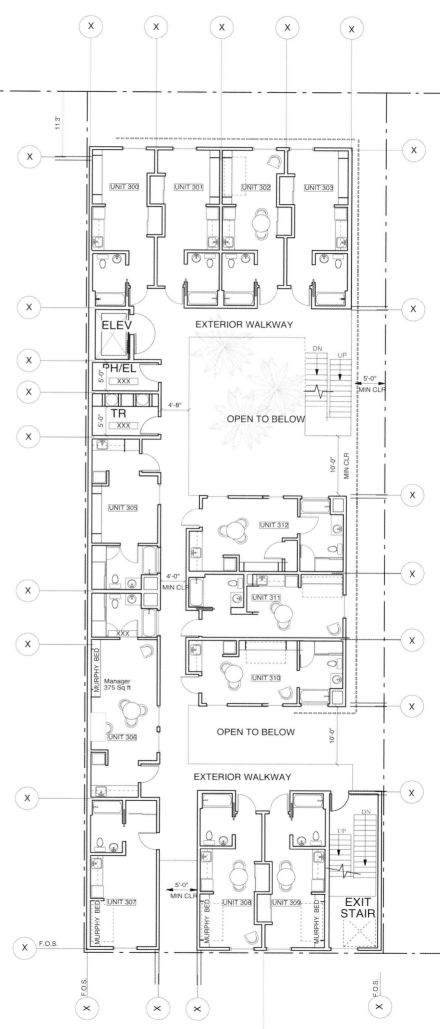

BAUFELD 10

BAUFELD 10 公寓

Architect: LOVE Architecture and Urbanism
Location: Harbour City, Hamburg, Germany
Site Area: 810 m²
Status: Completed
Units: 24
Photography: Anke Muellerklein

设计公司：LOVE 城市建筑设计事务所
地点：德国汉堡市港口城
占地面积：810 m²
状态：已建成
户数：24
摄影：Anke Muellerklein

The original construction cost is € 5,500,000
that means: construction cost in China is

¥ 9,781,120

per unit cost is ¥ 407,547

原建造成本为 5 500 000 欧元，
国内建造成本约

¥ 9 781 120 元

国内平均每户造价约 407 547 元。

With the HafenCity Hamburg, a new district covering 157 hectares is being developed directly at the port. In addition to the mixed utilization, the relevant urban development concept calls for high-quality architecture. For this reason, there was a separate tendering process for each individual building site.

LOVE architecture and urbanism from Graz won the competition for "Baufeld 10." The site is situated in an area within the Dallmannkai, directly on the water and in direct proximity to the "Elbphilharmonie" – a concert hall currently being developed by the Swiss Office of Herzog & de Meuron. In total, 26 architecutral offices were awarded projects at the Kaiserkai.

在汉堡的港口区，一个面积为 1 570 000 m² 的新街区直接建在了港口边上。除了具有综合用途，相关的城市开发概念还需要建造高质量的建筑。因此，每一座建筑都进行了独立招标。

来自格拉茨的 LOVE 城市建筑设计事务所赢得了 Baufeld 10 公寓的设计竞赛。项目基地位于 Dallmannkai 内的一片区域内，就处在水边，与 Elbphilharmonie 非常接近。Elbphilharmonie 是一座音乐厅，目前正由 Herzog & de Meuron 的瑞典办事处进行设计开发。共有 26 家建筑事务所在 Kaiserkai 获得了项目委托。

South Elevation

Cost and Creativity 成本与创意

Advanced energy saving techniques help the building be efficient in the future and only need maxium maintenace.

先进的节能技术能让建筑在未来同样保持高效，并且只需基本的维护。

The apartment building was planned and built according to the latest Energy Conservation Regulations (EnEV) and featured at energy-saving and climate protection.

公寓的规划和建设是根据最新的节能条例，具有节能和环保的功能。

LOVE's special challenge: The Baufeld 10 project was developed in a joint building venture. This means that the various future residents worked together to create a real community for the new building. Within this model, individualists connected with each other with the goal of building THEIR communal house. The building typology had to meet this expectation. This is why the building houses many different building typologies with all kinds of furnishing standards: from very large apartments (up to approx. 225 m²) to smaller units (approx. 50m²), which feature entirely different designs – from one-storey apartments to maisonettes that stretch across four storeys.

For Baufeld 10, individualisation was the top priority. Each of the 28 new residents can now enjoy his or her unique lifestyle within his or her apartment – whether horizontal or vertical – whether in a small or a big apartment. These different visions were blended into one building – to everyone's satisfaction.

LOVE 城市建筑设计事务将面临一个特殊的挑战：Baufeld 10 公寓是一个合资建筑项目。这意味着将来会由众多不同的住户为新建筑创造一个真正的社区。因此，大家共同合作，朝着建设他们的共同家园这一目标而努力。建筑类型必须满足这种期望，因此建筑内容纳了多种不同的类型，包括所有的配套标准——从超大的公寓（最大达 225 m²）到小公寓（50 m²），而且有着完全不同的设计——从单层公寓到四层的复式公寓。

个性化是 Baufeld 10，公寓设计最先考虑的因素。现在，28 个新住户中的每一个在自己的公寓内都可以享有自己的生活方式，无论公寓是水平的还是垂直的，无论公寓是大是小，这些不同的公寓融合在一座建筑内，满足了所有人的要求。

0   5   10   20

East Elevation

Viewed from the outside, the residential building presents a gleaming white structure formed of slightly bevelled cubes and with generously proportioned, slightly bevelled window openings. The configuration of these window openings matches the layout of the apartments behind them. Each apartment has a balcony and/or bay which protrude from the building. Within the rigid row of buildings, this configuration provides maximum views in an "exciting direction" – namely, the harbour.

The support system consists of a supporting building envelope with stiffening apartment partitions and storey ceilings and was calculated as a spacial static system. This made it possible to place the quite large window openings freely in the exterior walls of the building and to minimize the concrete and steel volume used, which also reduced the construction costs. More structurally demanding parts of the building, such as the roundings of the facade, the balconies and the bays, were executed as prefabricated elements. The facade, which serves as an outside composite thermal insulation system, was provided with a "natural stone plaster" surface. This is composed of natural stone grains with an admixture of mica, which causes the facade to sparkle slightly in the sun, a valuable visual effect. The window roundings and the bevelled window reveals were modelled into the composite thermal insulation system with thermal insulation moulding.

从外面看，公寓是一个由多个稍微倾斜的立方体组成的闪闪发光的白色结构，上面设置了稍微倾斜的大尺度开窗。这些窗洞的设置与其后的公寓布局十分匹配。每一间公寓都有一个阳台和从建筑凸出的凸窗。在排列紧密的建筑中，这样的设置方式能在理想方向中获得最大的视野，也就是朝向海港的视野。

公寓的支持系统包括与建筑配套的围护结构、坚实的阻断、天花，并将之作为一个空间的静态系统。这样方便在外墙上自由设置大型的窗口，并尽量减少混凝土和钢材的使用量，降低了建设成本。公寓其他有结构要求的部分，如立面的圆形部分、阳台和凸窗等，都是以预制部件打造而成的。立面作为复合外保温系统，具有"天然石膏"的表皮，里面掺杂有天然石粒和云母，让立面在阳光下闪闪发光，创造了别样的视觉效果。圆角的窗户和倾斜的窗框是在保温系统下以保温模型所制成的。

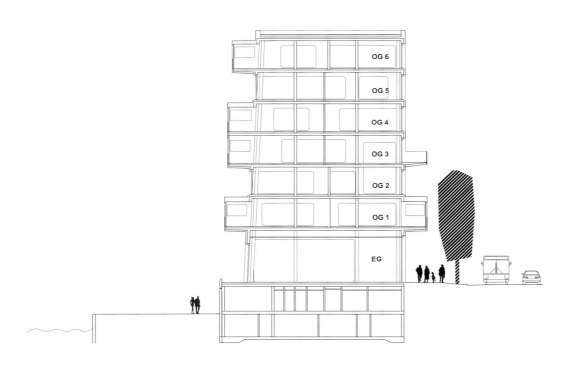

OG 6

OG 5

OG 4

OG 3

OG 2

OG 1

EG

0   5   10   20

Section

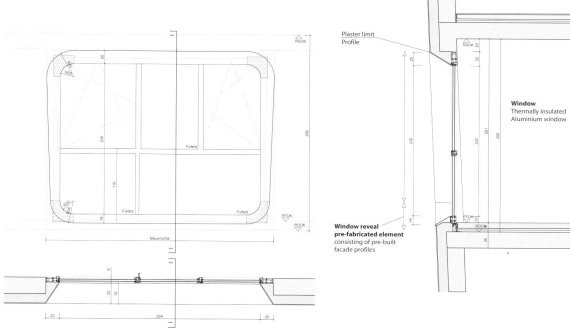

Window
Thermally insulated
Aluminium window

Plaster limit
Profile

Window reveal
pre-fabricated element
consisting of pre-built
facade profiles

Detail Section

From the planning stages of the new district, the Municipal Development Company already emphasized energy savings and climate protection with a series of sustainable ecological measures for the emerging new buildings.

Regarding the building's technical characteristics, the apartment building was planned and built according to the latest Energy Conservation Regulations (EnEV). The building features a very light construction mass and a building shell that provides excellent air sealing. One definite cornerstone of the building's energy concept is its connection to the local heat and power plant, which combines the production of heat and power. Outfitting the apartment building with a solar thermal system of ten vacuum-tube collectors for the central domestic water supply is an additional energy efficiency measure.

The solar thermal energy system and the heating connection to the local energy provider combine to cover the heating requirement of about 30 kwh/(m²a).

从新区的规划阶段，城市发展公司已经强调节约能源和气候保护的重要性，并要求新建筑执行一系列的可持续发展生态措施。

关于建筑的技术特点，公寓楼规划和建设是根据最新的节能条例（EnEV）。 该建筑采用了很轻的建筑质量和建筑外壳，可提供极佳的空气密封。 其中一个建筑的能源概念明确的基石连接到当地热电厂，它结合了热量和电力生产。 舾装用太阳能十真空管集热器热系统，中央生活供水公寓建设是一项额外的提高能源效率的措施。

太阳热能系统由本地能源供应商提供，功率为 30 kW·h/（m²·a）。

30 DEGREES SOUTH

30° 朝南公寓

Architect: PUSHAK Architects
Client: Municipality of Oslo
Location: Rommen, Oslo, Norway
Built Area: 18,000 m²
Status: Completed
Units: 200

设计公司：PUSHAK 建筑事务所
客户：奥斯陆市政府
地点：挪威奥斯陆市 Rommen
建筑面积：18 000 m²
状态：已建成
户数：200

The overall scheme is based on an existing masterplan that gave two rows of housing surrounding a shared green space. Within these frames the houses are angled in plan and section to optimize use of solar energy. The roofs have solar collectors and the facades have large windows towards the south. Greenery is used for solar shading in summer.

项目的整体方案建立在已规划好的总纲上：两排房屋围绕着一片公共绿地。根据已定的框架，公寓楼以一定角度的朝向获取和优化利用太阳能。屋顶有太阳能集热器，立面有朝南的大型窗户。丰富的绿化可以阻挡夏日的烈阳。

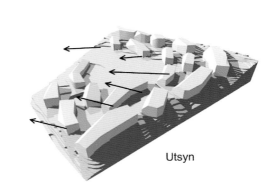

Utsyn

Solfangere 45° mot sør og +/-30° i plan.

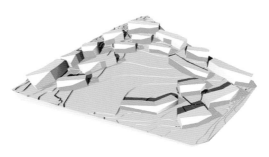

Sydvendte fasader +/- 30° sør

Cost and Creativity 成本与创意

Using solar collectors instead of PV panels is cost effective. Wooden facades will give low building costs in Norway.

在光伏电池板上安装太阳能集热器，是控制成本效益的一个重要部分。在挪威，木制的立面也有助于减少成本。

The development will have passive house standard, with minimized use of energy.

本项目完全遵照房屋建设标准，并将能源的消耗降到最低。

SUSTAINABLE TOWN HOUSES

可持续的城市住宅

Architect: C. F. Møller Architects,
Berg Arkitektkontor
Client: Byggnadsfirman Erik Wallin AB
Location: Stockholm, Sweden
Site Area: 3,200 m²
Status: Under Construction

设计公司：C. F. Møller 建筑事务所、
Berg Arkitektkontor
客户：Byggnadsfirman Erik Wallin AB
地点：瑞典斯德哥尔摩
占地面积：3 200 m²
状态：建设中

The original construction cost is € 9,000,000
that means: construction cost in China is

¥ 15,016,073

原建造成本为 9 000 000 欧元，
国内建造成本约

¥ 15 016 073 元

Hammarby Sjöstad in Stockholm is known worldwide as an exemplary example of how a former industrial port area can be transformed into a sustainable urban development. Now Stockholm is planning yet another high-profile environmental area on the harbourfront called Norra Djurgaardsstaden. Here, C. F. Møller Architects has won the competition for a town house development.

斯德哥尔摩的哈尔博（Hammary）因成功将老港口工业区改造成可持续发展的城市开发区而举世闻名。如今斯德哥尔摩正在 Norra 规划着另一个开发区，C. F. Møller 建筑事务所赢得了市政厅开发的竞争。

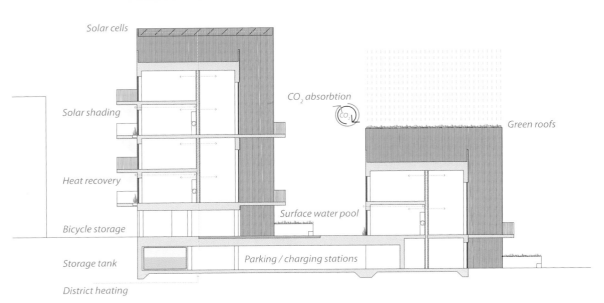

Solar cells

Solar shading

CO_2 absorbtion

Green roofs

Heat recovery

Bicycle storage

Surface water pool

Storage tank

Parking / charging stations

District heating

Cost and Creativity 成本与创意

The project makes it possible to live a modern life based on sustainable solutions. A rational pre-fabricated wood construction is combined with recyclable cladding and state-of-the-art low-energy solutions, to create a sustainable approach that is continuous throughout the building life cycle.

该项目以可持续发展方案为住户提供现代的生活，合理的预制木构结构结合可回收面板，加上国家最先进的低能源方案，打造出了这栋终身可持续发展的建筑。

The project makes it possible to live a modern life based on sustainable solutions.

项目为人们实现可持续发展的现代化生活创造了可能。

The town houses, 18 in total, will be neighbouring a former gasworks, which will house the district's cultural centre, as well as Husarviken, which flows into the archipelago, and the Stockholm National City Park. The architecture is inspired by the gasworks' red bricks and simple geometry and by the area's green qualities.

The residences, five different types all in all, will be built as staggered modules, a solution creating unique homes with rich daylighting and splendid views of the waterside. The design also gives room to private outdoor spaces and terraces.

新市政厅总计18层，邻近一座废煤气厂，它涵盖行政区的文化中心和斯德哥尔摩国际城市公园，其建筑风格从煤气厂的红砖、简单的几何形状和该区域绿色品质中得到启发。

住宅有五种类型，像交错搭建的模块，使其有充足的日光和极好的河畔视野，房间有私人空间和阳台。

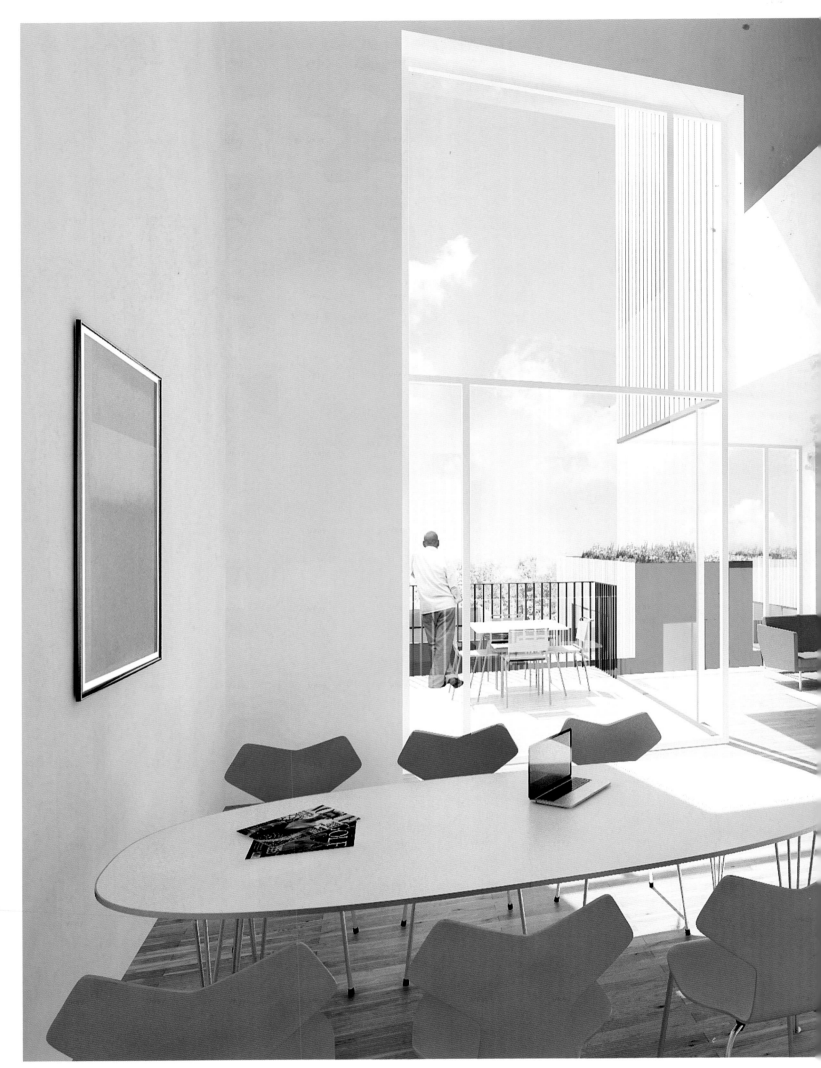

The aim of the Norra Djurgaardsstaden housing district is to adapt to global climate changes, so that in 2030 the district will no longer make use of fossil fuels - and thereby not contribute to the emission of CO2. The energy consumption of the living units will not exceed 55 kWh/m²/year, including a maximum of 15 kWh/ m²/ year used for electricity.

The energy consumption will be minimized by means of the massing of the buildings, for example, the town houses are staggered in order to maximize daylight and have dense constructions. Also contributing are solutions such as intelligent lighting, solar panels for heating, and heat recovery.

Norre 的目标是使住宅适应全球气候和变化。到 2030 年，该地区不再使用化石燃料，不再排放二氧化碳，生活能源消耗不超过每年 55 kW·h/m²，其中还包括 15 kW·h 的电力消耗。

大量建筑物对能源的消耗会得到重视，市政厅大楼也将密集交叠，以最大限度地利用日光，智能照明、太阳能制热和热回收利用等节能措施都得到应用。